Think
like a
MATHEMATICIAN

Think like a MATHEMATICIAN

Get to grips with the language of numbers and patterns

Anne Rooney

This edition published in 2019 by Arcturus Publishing Limited
26/27 Bickels Yard, 151–153 Bermondsey Street,
London SE1 3HA

AD006839UK

Printed in the UK

CONTENTS

What is mathematics, really?

Mathematics is all around us. It is the language that lets us work with numbers, patterns, processes and the rules that govern the universe. It provides a way for us to understand our surroundings, and both model and predict phenomena. The earliest human societies began to investigate mathematics as they tried to track the movements of the Sun, Moon and planets, and to construct buildings, count flocks and develop trade. From Ancient China, Mesopotamia, Ancient Egypt, Greece and India, mathematical thought flowered as people discovered the beauty and wonder of the patterns that numbers make.

Mathematics is a global enterprise and an international language. Today, it underlies all areas of life.

Trade and commerce are built on numbers. The computers that are integral to all aspects of society run on numbers. Much of the information we are presented with on a daily basis is mathematical. Without a basic understanding of numbers and mathematics, it's impossible to tell the time, plan a schedule or even follow a recipe. But that's not all. If you don't understand mathematical information, you can be deceived and misled – or you might simply miss out.

Mathematics can be commandeered for both honourable and nefarious purposes. Numbers can be used to illuminate, explain and clarify – but also to lie, obfuscate and confuse. It's good to be able to see what's going on.

Computers have made mathematics a lot easier by making possible some calculations that could never have been achieved before. You will meet examples of this later in the book. For example, pi (symbol π, which defines the mathematical relationship between the circumference of a circle and its radius) can now be calculated to millions of places using computers. Prime numbers (which are only divisible by one and themselves) are now listed in their millions, again thanks to computers. But in some ways computers could be making mathematics less logically rigorous.

PURE AND APPLIED MATHEMATICS

Most of the mathematics in this book falls under the heading of 'applied mathematics' – it's mathematics that is being used to solve real-world problems, applied to practical situations in the world, such as how much interest is charged on a loan, or how to measure time or a piece of string. There is another type of mathematics which preoccupies many professional mathematicians, and that is 'pure' mathematics. It is pursued regardless of whether it will ever have a practical application, to explore where logic can take us and to understand mathematics for its own sake.

Now that it's possible to process very large amounts of data, far more reliable information can be extracted from empirical data (that is, data that can be directly observed) than ever before. This means that more of our conclusions can be – apparently safely – based on looking at stuff rather than working stuff out. For instance, we could examine lots and lots of data about weather and then make predictions based on what has happened in the past. We would not need any understanding of weather systems to do this, it would just work from what has been observed before on the assumption that – whatever forces lie behind it – the same will happen in the future with a certain degree of probability. It might well work, but that's not really science or mathematics.

Look first or think first?

There are two fundamentally different ways of working with data and knowledge, and so of coming up with mathematical ideas. One starts from thinking and logic, and the other starts from observations.

Think first: Deduction is the process of reasoning through logic using specific statements to produce predictions about individual cases. An example would be starting with the statement that all children have (or once had) parents, and the fact that Sophie is a child, to deduce that Sophie must therefore have (or have once had) parents. As long as the two original statements are verified and the logic is sound, the prediction will be accurate.

Look first: Induction is the process of inferring general information from specific instances. If we looked at a lot of swans and found they were all white we might infer from this (as people once did) that all swans must be white. But this is not robust – it just means we haven't yet seen a swan that is not white (see Chapter 10).

Being right and being wrong

Mathematicians are not always right, whether they begin with inductive or deductive methods. On the whole, though, deduction is more reliable and has been enshrined in pure mathematics since its origins with the Greek mathematician Euclid of Alexandria.

How it can go wrong

Our ancestors thought the Sun orbited the Earth, rather than the other way round. How would the movement of the Sun appear if it did go around the Earth? The answer is: exactly the same.

The model of the universe constructed by the Ancient Greek astronomer Claudius Ptolemy (c.AD90–168) accounted for the apparent movements of the Sun, Moon and planets across the sky. This was an inductive method: Ptolemy looked at the empirical evidence (what he observed for himself) and constructed a model to fit it.

As it became possible to make more accurate measurements of the movements of the planets, medieval and Renaissance astronomers devised ever more complex refinements to the mathematics of Ptolemy's Earth-centred model of the universe to make it fit their observations. The whole system became a horrible tangle as bits were added incrementally to explain every new observation.

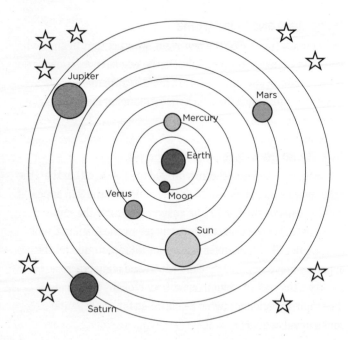

Putting it right

It was only when the model was overthrown in 1543 by the Polish astronomer and mathematician Nicolaus Copernicus, who put the Sun at the centre of the solar system, that the mathematics started to work. But even his calculations were not totally accurate. Later, the English scientist Isaac Newton (1642–1726) improved on Copernicus's ideas to give a mathematically coherent account of the movements of the planets which doesn't need lots of fudging to make it work. His laws of planetary motion have been validated by the observation of planets not discovered when he was alive. They have accurately predicted the existence of planets even before they were observed. But the model is not yet perfect; we still can't quite account for the motion of the outer planets, using our current mathematical model. There is more to be discovered, both in space and in mathematics.

Zeno's paradoxes

The mismatch between the world we experience and the world modelled by mathematics and logic has long been recognized.

The Greek philosopher Zeno of Elea (c.490–430BC) used logic to demonstrate the impossibility of motion. His 'paradox of the arrow' states that at any instant of time, an arrow is in a fixed position. We can take millions of snapshots of the arrow in all its positions between leaving the bow and reaching its target, and in any infinitely short instant of time it is motionless. So when does it move?

Another example is the paradox of Achilles and the tortoise. If the speedy Greek hero Achilles gave a tortoise a head start in a race, he would never be able to catch up with it. In the time it took Achilles to cover the distance to the tortoise's original position, the tortoise would have moved on. This would keep happening, with the tortoise covering ever-shorter distances as Achilles approached, but Achilles would never manage to overtake it.

This paradox works by treating the continuity of time and distance as a string of infinitesimal moments or positions. Logically coherent, it doesn't match reality as we experience it.

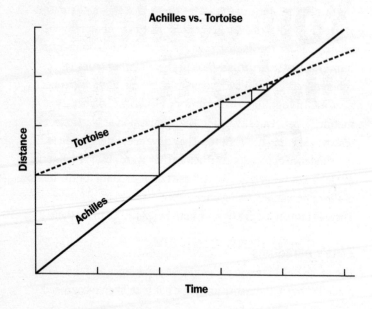

Achilles vs. Tortoise

CHAPTER 1

You couldn't make it up – or did we?

Is mathematics just 'out there', waiting to be discovered? Or have we made it up entirely?

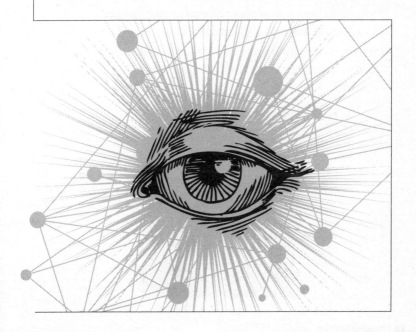

Whether mathematics is discovered or invented has been debated since the time of the Greek philosopher Pythagoras, in the 5th century BC.

Two positions – if you believe in 'two'

The first position states that all the laws of mathematics, all the equations we use to describe and predict phenomena, exist independently of human intellect. This means that a triangle is an independent entity and its angles actually do add up to 180°. Mathematics would exist even if humans had never come along, and will continue to exist long after we have gone. The Italian mathematician and astronomer Galileo shared this view, that mathematics is 'true'.

> *'Mathematics is the language in which God has written the universe.'*
>
> Galileo Galilei

It's there, but we can't quite see it

The Ancient Greek philosopher and mathematician Plato proposed in the early 4th century BC that everything we experience through our senses is an imperfect copy of a theoretical ideal. This means every dog, every tree, every act

of charity, is a slightly shabby or limited version of the ideal, 'essential' dog, tree or act of charity. As humans, we can't see the ideals – which Plato called 'forms' – but only the examples that we encounter in everyday 'reality'. The world around us is ever-changing and flawed, but the realm of forms is perfect and unchanging. Mathematics, according to Plato, inhabits the realm of forms.

Although we can't see the world of forms directly, we can approach it through reason. Plato likened the reality we experience to the shadows cast on the wall of a cave by figures passing in front of a fire.

If you are in the cave, facing the wall (chained up so that you can't turn around, in Plato's scenario) the shadows are all you know, so you consider them to be reality. But in fact reality is represented by the figures near the fire and the shadows are a poor substitute. Plato considered mathematics to be part of eternal truth. Mathematical rules are 'out there' and can be discovered through reason. They regulate the universe, and our understanding of the universe relies on discovering them.

What if we made it up?

The other main position is that mathematics is the manifestation of our own attempts to understand and describe the world we see around us. In this view, the convention that the angles of a triangle add up to 180° is just that – a convention, like black shoes being considered more formal than mauve shoes. It is a convention because we defined the triangle, we defined the degree (and the idea of the degree), and we probably made up '180', too.

At least if mathematics is made up, there's less potential to be wrong. Just as we can't say that 'tree' is the wrong word for a tree, we couldn't say that made-up mathematics is wrong – though bad mathematics might not be up to the job.

Alien mathematics

Are we the only intelligent beings in the universe? Let's assume not, at least for a moment (see Chapter 18).

> **'God created the integers. All the rest is the work of Man.'**
>
> Leopold Kronecker
> (1823–91)

If mathematics is discovered, any aliens of a mathematical bent will discover the same mathematics that we use, which will make communication with them feasible. They might express it differently – using a different number base, for example (see Chapter 4) – but their mathematical system will describe the same rules as ours.

If we make up mathematics, there is no reason at all why any alien intelligence should come up with the same mathematics. Indeed, it would be rather a surprise if they did – perhaps as much of a surprise as if they turned out to speak Chinese, or Akkadian, or killer whale.

For if mathematics is simply a code we use to help us describe and work with the reality we observe, it is similar to language. There is nothing that makes the word 'tree' a true signifier for the object that is a tree. Aliens will have a different word for 'tree' when they see one. If there is nothing 'true' about the elliptical orbit of a planet, or about the mathematics of rocket science, an alien intelligence will probably have seen and described phenomena in very different terms.

How amazing!

Perhaps it is amazing that mathematics is such a good fit for the world around us – or perhaps it is inevitable. The 'it's amazing' argument doesn't really support either view. If we invented mathematics, we would create something that adequately describes the world around us. If we discovered mathematics, it would obviously be appropriate to the world around us as it would be 'right' in a way that is larger than us. Mathematics is 'so admirably appropriate to the objects of reality' either because it's true or because that's what it was designed for.

Look out – it's behind you!

Another possibility is that mathematics seems astonishingly good at representing the real world because we only look at the bits that work. It's rather like seeing coincidences as evidence of something supernatural going on. Yes, it's really amazing that you went abroad on holiday to an obscure village in Indonesia and bumped into a friend –

> *How can it be that mathematics, being after all a product of human thought which is independent of experience, is so admirably appropriate to the objects of reality?'*
>
> Albert Einstein
> (1879–1955)

but only because you are not thinking about all the times you and other people have gone somewhere and not bumped into anyone you knew. We only remark on the remarkable; unremarkable events go unnoticed. In the same way, no one thinks to fault mathematics because it can't describe the structure of dreams. So it would be reasonable to collate a list of areas where mathematics fails if we want to assess its level of success.

'The unreasonable effectiveness of mathematics'

If mathematics is made up, how can we explain the fact that some mathematics, developed without reference to real-world applications, has been found to account for real phenomena often decades or centuries after its formulation?

As the Hungarian-American mathematician Eugene Wigner pointed out in 1960, there are many examples of mathematics developed for one purpose – or for no purpose – that have later been found to describe features of the natural world with great accuracy. One example is knot theory. Mathematical knot theory involves the study of complex knot shapes in which the two ends are connected. It was developed in the 1770s, yet is now used to explain how the strands of DNA (the material of inheritance) unzip themselves to duplicate. There are still counter-arguments. We only see what we look for. We choose the things to explain, and choose those that can be explained with the tools we have.

Perhaps evolution has primed us to think mathematically and we can't help doing so.

Knot theory: the simplest possible true knot is the trefoil or overhand knot, in which the string crosses three times (3_1 below). There are no knots with fewer crossings. The number of knots increases rapidly thereafter.

> '**How do we know that, if we made a theory which focuses its attention on phenomena we disregard and disregards some of the phenomena now commanding our attention, that we could not build another theory which has little in common with the present one but which, nevertheless, explains just as many phenomena as the present theory?**'
>
> Reinhard Werner
> (b.1954)

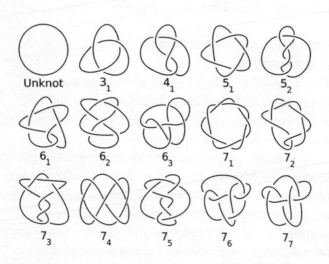

Unknot 3_1 4_1 5_1 5_2

6_1 6_2 6_3 7_1 7_2

7_3 7_4 7_5 7_6 7_7

Does it matter?

If you are just working out your household accounts or checking a restaurant bill, it doesn't much matter whether mathematics is discovered or invented. We operate within a consistent mathematical system – and it works. So we can, in effect, 'keep calm and carry on calculating'.

For pure mathematicians, the question is of philosophical rather than practical interest: are they dealing with the greatest mysteries that define the fabric of the universe? Or are they playing a game with a kind of language, trying to write the most elegant and eloquent poems that might describe the universe?

> '*The miracle of the appropriateness of the language of mathematics for the formulation of the laws of physics is a wonderful gift which we neither understand nor deserve.*'
>
> Eugene Wigner

Where the 'reality' of mathematics matters most is where humans are pushing against the boundaries of knowledge and of technical achievement. If mathematics is made up, we might come up against the limitations of our system and not be able to push through them to answer certain questions. We might never achieve time travel, zip to the other side of the universe, or create artificial consciousness, simply because our mathematics is not up to the task. We will deem impossible things which, with a different system of mathematics, might be perfectly easy.

On the other hand, if mathematics is discovered we can, potentially, uncover all of it and achieve right to the edges of what is possible, of what is allowed by the physical laws of the universe. It would be nice, then, if mathematics were discovered. But we can't be certain.

A DREADFUL POSSIBILITY

One possibility that doesn't usually get much consideration is that mathematics is real, but we've got it all wrong, just as Ptolemy got the model of the solar system wrong. What if the mathematics we have developed is the equivalent of the Ptolemaic Earth-centred universe? Could we throw it away and start again? It's hard to see how that would be possible now we have invested so much in it.

CHAPTER 2

Why do we have numbers at all?

Getting to grips with numbers came early in the development of human society.

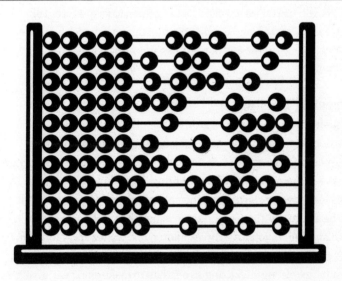

We are so used to numbers that we rarely give them a second thought. Children learn to count at a very early age, with numbers and colours being among the first abstract ideas they encounter.

Tally ho!

The first human engagement with numbers that we know of was in the form of tallying. Our distant ancestors kept tallies of their flocks by marking a stick, stone or bone, with one cut for each animal, or by moving pebbles or shells from one pile to another.

Tallying doesn't need words for the numbers – it's not the same as counting. It's a simple system of correspondence, using one object or mark to represent another object or phenomenon. If you have a shell that represents each sheep, and you drop a shell into the pot as each sheep passes, it's easy to see if you have shells left over, and so sheep missing, at the end. You don't need to know whether you should have 58 sheep or 79 sheep – you just keep looking for the missing sheep and dropping a shell in the pot each time one is found until there are no shells left over.

We still use tallies to keep score in games, to keep a record of days shipwrecked, and in other circumstances in which a number is only needed at the end of a process. Counting comes after tallying.

Counting 1, tallying 0

Tallying was used by various Stone Age cultures for at least 40,000 years. Then at some point it became useful to have numbers with names.

We don't know quite when counting began, but it's easy to see that once people started keeping animals it would be more useful to be able to say 'three sheep are missing' than just 'some sheep are missing'. If you have three children and want a spear for each, it's easier to know you have to make three spears, then set out to find three strong sticks, and so on, than to make one spear, give it to the first child, realize there are still spearless children, make another spear, and so on. Once people started to trade, numbers would have been essential.

The first known written numbers emerged in the Middle East in the Zagros region of Iran around 10,000BC. Clay tokens used in counting sheep have survived. The token for a single sheep was a ball of clay with a + sign scratched into it. Clearly that's great if you have a few sheep, but needing 100 tokens for 100 sheep would be cumbersome. They developed tokens with different symbols to represent 10 sheep and 100 sheep, and could then account for any number of sheep with far fewer tokens – even 999 sheep could be represented with only 27 tokens (9 x 100-sheep tokens; 9 x 10-sheep tokens; 9 x single-sheep tokens).

The tokens could be strung on a cord, or were often baked into a hollow clay ball. The outside of the ball was impressed with symbols showing the number of 'sheep' inside, but it could be broken to verify the number if there was a dispute. These numbers on the outside of sheep-counting balls are the oldest surviving written number system.

Making up numbers

Many early number systems developed directly from tallies and so used a symbol repeated for units, a different symbol for tens, and another for hundreds. Some had symbols for 5, or other intermediate numbers.

The system of Roman numerals, familiar from clockfaces and the copyright date shown at the end of a movie, began with the vertical strokes of a tallying system. The numbers 1–4 were originally represented as I, II, III, IIII. X is used for 10 and C for 100. The intermediates V (5), L (50) and D (500) make large numbers a bit shorter to write. After a while, a convention emerged of putting a I before a V or X to denote subtraction, so IV is 5 – 1, or 4. IV is shorter to write and easier to read than IIII. You can only do it within the same power of ten, so IX is 9 but you can't write IC for 99 – it has to be XCIX (or 100 – 10 and 10 – 1).

1	2	3	4	5	6	7	8	9	10
I	II	III	IIII later IV	V	VI	VII	VIII	VIIII later IX	X

11	19	20	40	50	88	99	100	149	150
XI	XIX	XX	XL	L	LXXXVIII	XCIX	C	CXLIX	CL

Limited by numbers

Using repeated symbols to stand for extra units, tens and hundreds, makes numbers cumbersome to write and makes arithmetic difficult. With a system, like the Roman one, of preceding a symbol with one to be subtracted, addition can't even be achieved by just counting up the total number of each type of symbol: XCIV + XXIX (94 + 29) would give the same answer as CXVI + XXXI (116 + 31) if we just counted Cs, Xs, Vs and Is. Although the Romans managed, the system has clear limitations:

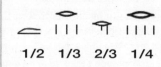
their mathematics was too inflexible. Fractions were all based on division by 12, there were no decimal fractions – and can you imagine trying to deal with complex concepts such as powers (see box on page 28) or quadratic equations using Roman numerals and with no figure for 0?

IV^{III} = LXIV
XIIx^{II} + IVx – IX = I – I

Not surprisingly, Roman mathematics didn't develop very far.

Place value

The Indo–Arabic numeral system we use today has only nine figures, which can be reused *ad infinitum*. It developed slowly in India from the 3rd century BC and was later refined by Arabian mathematicians before being adopted in Europe. In this system, the status of a number is indicated by its position, called *place value.* Place value increases moving towards the left. This is a much more flexible system than the Roman one.

Thousands	Hundreds	Tens	Units
5	6	9	1

We can make a number such as 5,691 by combining:

5,000	(5 x 1,000)
600	(6 x 100)
90	(9 x 10)
1	(1 x 1)

Using place value, it's possible to represent even very large numbers with a small number of figures. Compare the Roman and Arabic representations:

88	**= LXXXVIII**
797	**= DCCXCVII**
3,839	**= MMMDCCCXXXIX**

Nothing there – the start of zero

Place value is all very well as long as there is a digit in each place. If there are gaps – nothing in the 10s column (308, for instance) – how can we show this? Leaving a space, as the Chinese did, can be ambiguous unless the numbers line up carefully in columns: 9 2 could be 902 or possibly 9002, and there's a big difference between the two.

A space indicated an empty column in Indian numbers, too, but was later replaced by a dot or small circle. This was given the Sanskrit name *sunya*, meaning empty. When the Arabs adopted the Indian numerals, around AD800, they also took the empty place-marker, still calling it empty, which was *sifr* in Arabic and is the origin of the modern word 'zero'.

The earliest surviving use of a symbol for zero in decimal figures is a Cambodian inscription on stone dating from 683. The large dot stands for 0 between the figures for 6 and 5, denoting 605.

Indo–Arabic numerals first appeared in Europe around AD1000, but it was several centuries before they were universally adopted. The Italian mathematician Leonardo Bonacci, better known today as 'Fibonacci', promoted their use as early as the 1200s, yet merchants continued to use Roman numerals until the 16th century.

CHAPTER 3

How far can you go?

Not all number systems
are infinitely extendable.

Our number system is unlimited – it can go up to any number you care to imagine, just by putting down more and more digits. That has not always been the case.

Not enough numbers?

The simplest counting systems are called 2-count. They don't provide a way of doing calculations, but allow counting of small quantities. A 2-count system has words for 1, 2 and sometimes 'many' (meaning an uncountably large number). The 2-count system used by bushmen in South Africa builds in a series of 2s and 1s. Its usefulness is limited by how many 2s people can keep track of.

1	xa
2	t'oa
3	'quo
4	t'oa-t'oa
5	t'oa-t'oa-ta
6	t'oa-t'oa-t'oa

Supyire, a language spoken in Mali, has basic number-words for 1, 5, 10, 20, 80 and 400. The rest of the numbers are built up from these. For example, 600 is *kàmpwòò ná kwuu shuuní ná bééshùùnnì*, which means **400 + (80 x 2) + (20 x 2)**.

The Toba in Paraguay use a system which has words for numbers up to 4, and then starts reusing words extravagantly:

1	nathedac
2	cacayni or nivoca
3	cacaynilia
4	nalotapegat
5 = 2 + 3	nivoca cacaynilia

6 = 2 x 3	cacayni cacaynilia
7 = 1 + 2 x 3	nathedac cacayni cacaynilia
8 = 2 x 4	nivoca nalotapegat
9 = 2 x 4 + 1	nivoca nalotapegat nathedac
10 = 2 + 2 x 4	cacayni nivoca nalotapegat

This sort of system is fine for counting your children or other things that come in relatively small quantities, but it has clear limitations.

A small infinity

Infinity is often considered to be an uncountably large number (see Chapters 7 and 8). For the Toba and the South African bushmen using 2-count, that might well be a number below 100. In a society not concerned with abstract mathematics, there is no need to raise the bar for infinity much further than the size of a family or herd of animals.

Less than zero

In early run-of-the-mill counting, there was no need for negative numbers. Indeed, the Ancient Greeks were highly distrustful of them, and the mathematician Diophantus, in the 3rd century AD, said that an equation such as 4x + 20 = 0 (which is solved with a negative value for x) is absurd.

A TAXONOMY OF NUMBERS

Mathematicians now recognize several categories of numbers.

- *Natural* numbers are those you first learn about, the numbers we count with: 1, 2, 3, and so on.

- *Whole* numbers are the natural numbers with zero chucked in: 0, 1, 2, 3, and so on. (This might seem a bit odd, as how whole is zero? It's a lack of a number, a hole rather than a whole. Never mind, that's mathematicians for you.)

- *Integers* are whole numbers and the numbers below zero, the negative numbers: . . . -3, -2, -1, 0, 1, 2, 3 . . .

- *Rational* or *fractional* numbers are numbers that can be written as fractions, such as ½, ⅓, and so on. They include the integers as they can be written as fractions: ¹/₁, ²/₁, etc. They include all the fractions between whole numbers, as they can be written as fractions, too: 1 ½ can be written as ³/₂, and so on. All rational numbers can be written as either terminating or repeating decimals. So ½ is 0.5 and ⅓ is 0.33333 . . .

- *Irrational* numbers are those which can't be written as terminating or repeating decimals or expressed as a ratio between two whole numbers. They are decimals that go on and on in a non-repeating sequence. Examples are π, $\sqrt{2}$, and e, which can be calculated by computer to trillions of places without revealing a repeating pattern.

- *Real* numbers: all of the above.

- *Imaginary* numbers: numbers that include *i*, defined as the square root of -1. (We won't worry about that one.)

Certainly the early, tallying farmer who noticed that three sheep were missing did not need to say he or she had -3 sheep; it was good enough to say they were three short of a full flock. With commerce, though, came a need to show a debt. If you borrowed 100 coins, your account stood at -100; if you paid back 50 of them, your account stood at -50. Negative numbers were used for this purpose in India from the 7th century AD.

The first known appearance of negative numbers is even earlier. The Chinese mathematician Liu Hui established rules for arithmetic using negative numbers in the 3rd century. He used counting rods in two colours, one for gains and one for losses, which he called positive and negative. He used red counting rods for positive numbers and black for negative numbers – the opposite of the modern accounting convention.

Counting and measuring

While many things can be counted, not all can be counted easily and some can't be counted at all. In nature, there are perhaps more things that can't easily be counted than can be.

We can count people, animals, plants and small numbers of stones or seeds. But although in theory we could count the grains of wheat in a harvest or the number of trees in a forest or ants in an anthill, it's unlikely that we would. These are things we are likely to measure instead. Humans began measuring grain by weight or volume long ago. Some things can only be measured in this way: we measure the volume of liquids, the weight (or mass) of rocks and the area of land (see Chapter 15).

Further still from counting are the arbitrary scales for measurements such as temperature. Scales provide another use for negative numbers. Unless a scale starts at some form of absolute zero, a negative number can be useful. Thermometers most certainly need negative numbers, if working in Celsius or even Fahrenheit. Negative numbers are needed with vectors (a quantity that also includes direction), as we express one direction as positive and the opposite as negative. If we turn clockwise through 45°, that is a positive rotation, but if we then turn back 30°, that's a rotation of -30°. Ions (electrically charged particles) can have a positive or negative charge, and which charge they have indicates how they will react with other substances. You might come across negative numbers on a daily basis in circumstances such as:

- Floor -1 in a lift – a floor below ground level, which is considered to be 0

- A soccer club with a negative goal difference – more goals conceded than scored

- A negative altitude, indicating that a geographical location is below sea level

- Negative inflation (deflation) showing that retail prices are dropping.

Who counts?

Although we think of mathematics as a uniquely human activity, some other animals seem to be able to count. Scientists have found that some types of salamander and fish can distinguish between different sized groups as long as the ratio of one to the other is greater than two. Honeybees can apparently distinguish numbers up to four. Lemurs and some types of monkey have limited numerical abilities, and some types of bird can count well enough to know if their eggs or chicks are missing.

This sort of system is fine for counting things that come in relatively small quantities, but it has clear limitations.

ARE NUMBERS REAL?

Of all the candidates for reality in mathematics, the whole numbers seem to have the best claim. Even the Polish mathematician Leopold Kronecker accepted them.

Whole numbers seem quite healthy until you look closely, as though they could be found in nature. Perhaps three wolves run through the forest. That's an event in the natural world which looks as though it works with whole numbers. But we can't actually put a rigid boundary around each wolf. There are always atoms flying off the wolf, moving in and out of it; it's picking up more electrons from getting a static charge by rubbing against another wolf; even most of its cells are not actually bits of wolf. There is an entity that is approximately one wolf, but it's ever-changing. We can go smaller and smaller, down to subatomic particles, and even then we find a 'thing' is a cloud or pulse of energy that might or might not be in a particular position at any moment. Hard to count.

Are whole numbers a snapshot of a moment? How short is the moment? How are we measuring it? The measurement of a continuity such as time is entirely arbitrary. And, as Zeno's paradoxes show (see page 13), if we break time into ever shorter moments the logical results don't match the reality we observe.

CHAPTER 4

How many is 10?

Ten is generally considered to be one more than nine – but it doesn't have to be.

We say our number system uses base-10, which means that when we get as far as nine, we start again with 0 in the units column and 1 in the next column, which we designate 'tens'. Succeeding numbers use two digits, one showing the tens and one showing the units. When we get to 99, we've run out of digits we can put in both places and start another column, for hundreds.

It doesn't have to be this way – there is no rule that says 9 has to be the highest digit we can put in a column. We could use more or fewer digits.

What is base 10?

The name 'base 10' tells us nothing; at whichever number we stop counting units, the first number using a new column is always going to be '10'. An alien race that counts in base 9 will also call their system base-10 and will have no digit for, say, '9' (0, 1, 2, 3, 4, 5, 6, 7, 8, 10). We really need a new name (and squiggle) for the '10' we use just to name the base.

Fingers, toes, legs and tentacles

We have probably developed a base-10 number system because we have ten fingers and thumbs, so that makes counting in tens easy. If, instead of humans, three-toed sloths had become the dominant species, perhaps they would have developed a base-6 or base-3

number system – or even base-12 if they were happy to use the toes on their hind limbs as well as those on their forelimbs. A base-3 system would count like this:

Base 3 – counting in sloth #1									
0	1	2	10	11	12	20	21	22	100
Base 6 – counting in sloth #2									
0	1	2	3	4	5	10	11	12	13
Base 10 – counting in human									
0	1	2	3	4	5	6	7	8	9

If octopuses had become the dominant species, they might have counted in base-8 (octal). In fact, as they are very intelligent creatures, they might well count in base-8 for all we know.

Base 8 – counting in octopus									
0	1	2	3	4	5	6	7	10	11
Base 10 – counting in human									
0	1	2	3	4	5	6	7	8	9

10, 20, 60 . . .

We don't even need to switch species to see different bases at work. The Babylonians worked in base 60 (see Chapter 6) and the Mayans used base 20.

Two-count systems use base-2 (see page 42). We have used base 12 as the basis for quite a few systems of measurement (12 inches in a foot, 12 pennies in an old shilling, 12 eggs in a dozen). Starting with the human body doesn't mean we have to end up with base 10, either.

The Oksapmin of New Guinea use base 27, derived from counting body parts starting with the thumb of one hand and moving up the arm to the face and down the other side to the opposite hand (see image below).

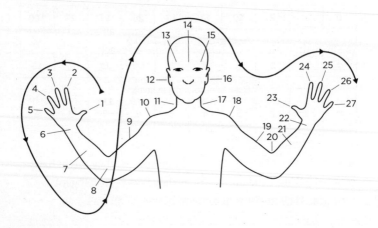

Computer counting

We don't use base-10 for everything. Many computing tasks use base-16, called hexadecimal. As we don't have any digits for numbers above 9, the letters at the start of the alphabet are co-opted to stand for the numbers from 10 to 15 in hexadecimal.

Base 10 – counting in human																
0	1	2	3	4	5	6	7	8	9	10	11	12	13	14	15	16
Base 16 – counting in computer #1																
0	1	2	3	4	5	6	7	8	9	A	B	C	D	E	F	10

You might have noticed codes such as #a712bb labelling colours on the computer. These are triplets of hexadecimal numbers – a7, 12, bb – which give a value for each of the three principal colours – red, green and blue – from which all other colours are built on a computer. These numbers, if converted to decimal (base-10) would be 23 (a7=16+7); 18 (12=16+2); and 191(bb=(11x16)+15). Using hexadecimal means that larger numbers (up to 255=ff) can be stored using only two digits.

Ultimately, all operations on a computer are reduced to binary, or base-2. This uses only two digits – 0 and 1 – as counting starts again with a new place every time we reach 2.

Base 2 – counting in computer #2									
0	1	10	11	100	101	110	111	1000	1001
Base 10 – counting in human									
0	1	2	3	4	5	6	7	8	9

Binary allows all numbers to be represented by one of two states, on/off or positive/negative. It means that anything can be coded on a magnetic disk or tape by the presence or absence of a charge.

Alien alert

If there are intelligent beings anywhere else in the universe, which seems quite possible (see Chapter 18), how would they count? They might have 17 tentacles and count in base-17. It is highly likely, though, that at some point they will have discovered and used binary (assuming numbers are not just a human construct). It could be that binary is the way we will be able to communicate with them.

The plaques fixed to the outside of the *Pioneer* spacecraft (see image on page 44) launched in 1972 and 1973 showed the binary states of hydrogen, with electron spin up and down. The difference between the two is used as a measure of time and distance and, being the same everywhere in the universe, should be recognized by a civilization capable of space travel

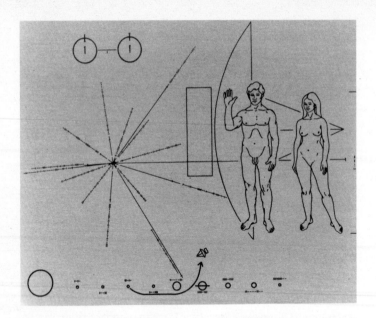

All your bases . . .

Is there another way to count? It might seem intuitive to take
the discrete numbers as the basis of our number system, but
there could be other ways of approaching numbers. What
if we counted in base pi and had a culture that was focused
on circles? What if our system were based on powers? This
is not entirely stupid, as it would focus on the difference
between one, two and three dimensional entities – lines, areas
and volumes. It's almost impossible for us to imagine how
these systems could work – but it's similarly impossible to
imagine how the world would look if a different part of the
electromagnetic spectrum were visible to us. Honeybees can
see ultraviolet light, for example, while rattlesnakes can see
in infrared. We can't rule out the possibility that different life
forms elsewhere in the universe might use numbers in an
entirely different way, or not at all.

Making bases work hard: logarithms

A logarithm is 'the exponent you have to raise a base to in order to get a specific number'. That sounds confusing, but it's not that hard. The expression:

$$y = b^x \Leftrightarrow x = \log_b(y)$$

(don't panic)

means, using some numbers as examples,

$$1000 = 10^3, \text{ so } \log_{10}(1000) = 3$$

Logarithms are a good way of dealing with very large numbers as they reduce them to much smaller numbers. To multiply numbers, add their logs together. To divide numbers, subtract one log from the other. Then de-logify the answer.

Before we used calculators and computers on a daily basis, tables of logarithms were the way to carry out complex calculations.

Fractional power

What is a little harder to grasp is that a number can be raised to a fractional power – that is, to a power that is not a whole number. The log in base 10 of 2, $\log_{10}(2)$, is 0.30103, which means that $10^{0.30103}$ = 2. How can a number be multiplied by itself less than a whole number of times?

Mathematics is a tricksy thing.

You could draw a graph of the powers of 2, and it would look like the one on page 46. (This is called a logarithmic curve, and lots of graphs follow this shape. The line approaches but never reaches the y-axis (x=0).)

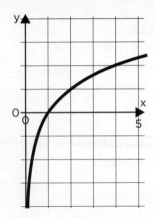

Once you've drawn a graph, you can read off any values you like, including the apparently impossible values of a number raised to a fractional power.

All logarithmic graphs, no matter what the base of the logarithms, cross the x-axis at 1 as any number raised to a power of zero is 1:

$$10^0 = 1$$
$$2^0 = 1$$
$$15.67^0 = 1$$

Clearly, the numbers go below 0, too. Negative powers yield values less than one as the minus sign tells us to put 1 over the number (the reciprocal of the number), making a fraction:

$$2^{-1} = \tfrac{1}{2}$$
$$2^{-2} = \tfrac{1}{2}^2 = \tfrac{1}{4}$$

And just in case you thought logarithms have to be in base 10 – they don't. For example, the logarithm in base 2 of 16 is 4:

$$16 = 2^4, \text{ so } 4 = \log_2(16)$$

A lot of science, engineering and even financial applications use so-called 'natural logarithms'. These are logarithms to the base e, which is an irrational number (a number with an unending decimal fraction) that starts 2.718281828459 . . .

All about e

The number called 'e', or Euler's number, is defined by mathematicians with this scary-looking expression:

$$e = \sum_{n=0}^{\infty} \frac{1}{n!}$$

It's actually pretty straightforward. All this means is:

$$e = 1 + \frac{1}{1} + \frac{1}{1x2} + \frac{1}{1x2x3} + \frac{1}{1x2x3x4} \ \cdot \cdot \cdot$$

and so on, up to infinity. The start of the sequence evaluates to:

$$e = \frac{1}{1} + \frac{1}{2} + \frac{1}{6} + \frac{1}{24}$$

$$= 1 + 1 + 0.5 + 0.1666 \ldots + 0.4166 \ldots$$

$$= 2.70826 \ldots$$

A natural logarithm is shown as $\log_e$ or ln. So $\log_e(n)$ is the power to which e must be raised to give the number n:

$$e^{1.6094} = 5$$

so

$$\log_e(5) = 1.6094$$

It might look useless, but it's used a lot to work out things such as compound interest. The formula for calculating compound interest on a deposit of 1 dollar/pound/euro at a rate R of annual interest over t years is e^{Rt}. If you invested your money for five years at 4 per cent interest, after five years you would have $e^{0.04 \times 5} = e^{0.2} = 1.22$. If you invested 10 dollars/pounds/euros, it would be:

$$10e^{0.04 \times 5} = 10e^{0.2} = 12.21$$

(The extra 0.01 is just the next digit in the answer, which goes to more decimal points than we can use with currency.

USES OF e: GET A JOB!

In 2007, Google put up posters in some American cities showing:

'{first 10-digit prime found in consecutive digits of e}.com'

Solving the problem and entering the web address (7427466391. com) led to an even more difficult problem, and solving that led to the Google Labs page, which invited the visiting geek to apply for a job.

CHAPTER 5

Why are simple questions so hard to answer?

Questions are easy to ask but difficult to answer, if you want rock-solid proof.

Can every whole even number be expressed as the sum of two primes? This question – which doesn't look as though it's very important to us in everyday life – seems deceptively simple. A Prussian amateur mathematician, Christian Goldbach, suspected that every whole number greater than 2 can be expressed as the sum of two prime numbers. In 1742 he wrote to the internationally-renowned mathematician Leonhard Euler to propose it. It's easy enough to try out a few numbers and see that it appears to work:

4 = 2 + 2 (2 is the only even prime)
6 = 3 + 3
8 = 5 + 3
10 = 5 + 5
12 = 7 + 5

and so on, to

7,614 = 7,607 + 7

and onwards ...

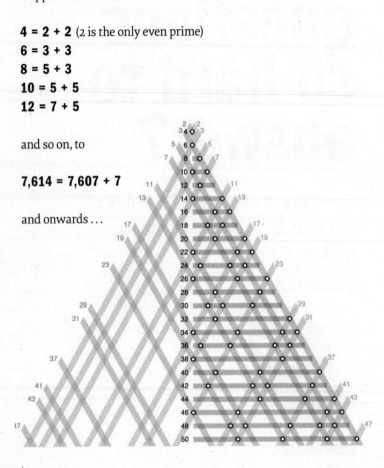

First and prime?

Although 'first' and 'prime' are synonyms in some contexts, the number 1 is not actually considered a prime number. The definition of prime numbers excludes it: 'any number greater than 1 that has no factors besides itself and 1.' There are other reasons, that are increasingly complex, but let's just take it as read that 1 is not a prime because it's too special.

In fact, Goldbach *did* consider 1 to be a prime. He had a second idea, now called the weak Goldbach conjecture, which stated that every odd number greater than 2 could be expressed as the sum of three primes. That has had to be rephrased to say every odd whole number greater than 5, so that we don't have to co-opt 1 into a role it's no longer allowed to occupy. (The weak conjecture was proven by the Peruvian mathematician Harald Helfgott in 2013.)

Euler, unwisely, was rather dismissive of Goldbach's idea. As it turned out, although Goldbach could try it out with a lot of numbers and it held up, he could not prove it. In mathematics, it's really not good enough for something to work with every number you try it with – there has to be a proof.

Goldbach's conjecture remains unproven to this day. Computers have tested it up to 4×10^{18} (4,000,000,000,000,000,000) – but that's still not good enough. What if there were a value, somewhere around $10^{2,000,000}$, for which it wasn't true? We would have been fooled into crediting it as a theorem when it wasn't. And even though the number $10^{2,000,000}$ is of no practical use, since there is not that number of anything in the known universe, it matters. Although just trying it out can never be enough to prove it, it could disprove it (see Chapter 10). For this reason, repeated trials are not wasted effort.

It's all conjecture . . .

In mathematics, a theorem is a statement that can be proven. If you don't have a proof for your idea – the idea might be a guess, a hunch, something backed up by lots of examples, perhaps – you can only ever claim it as a conjecture. If you later find a proof, you can upgrade it to a theorem. If someone else finds a proof, they usually get to name the theorem, even if it was thought of centuries before.

Fermat managed a nifty trick with his so-called 'last theorem' (see box below) in that he said he had a proof, but no space to write it down. When a proof was finally discovered by the English mathematician Andrew Wiles in 1993, the name 'Fermat's last theorem' continued to be used as Fermat claimed to have had a proof for it (and, in any event, it had become really famous with that name).

Who knows whether he did have a proof? Perhaps he didn't want it to be only a conjecture.

FERMAT'S LAST THEOREM

In 1637, Pierre de Fermat scribbled his 'last theorem' in the margin of a copy of *Arithmetica* by the Greek mathematician Diophantus. It states that no three integers a, b, and c (not 0) can satisfy the equation $a^n + b^n = c^n$ for any integer value of n greater than 2.

This means that while we can write, say,
$3^2 + 4^2 = 5^2$ (9 + 16 = 25),
we can't do the same for any powers greater than 2. Fermat noted that he had a proof, but it didn't fit in the margin so he had not written it down.

Can you prove it?

Simple questions in mathematics can be very hard to answer because of the difficulty of providing a proof. Goldbach said he was certain his idea was true, but he couldn't prove it. Computers could demonstrate that it is true for all useful numbers and a good many numbers way beyond being useful.

> 'That . . . every even integer is a sum of two primes, I regard as a completely certain theorem, although I cannot prove it.'
>
> Christian Goldbach,
> letter to Euler
> (7 June 1742)

A proof, in mathematics, is an inductive argument (as opposed to deduction). It must be based on other proofs that are already established (theorems) or statements that are self-evidently true, known as axioms. Proof, then, is based on logic and reasoning. Every step in a proof must be based on known truths. Very occasionally, if it is possible to examine every possibility, a proof can be based on examination of cases.

For instance, if we had a conjecture that applied to all even numbers between 2 and 400, we could examine each number in turn and see if it met the conditions. If it did, then we would have proved the conjecture and would have a theorem – but that's not normally the case. We can't, with reference to Goldbach's conjecture, check every even number since there is an infinite number of them. Instead we need a proof in which variables can stand in for numbers.

Euclid and those axioms

We rather glossed over those 'self-evident truths', or axioms. What makes something a self-evident truth? To you or me, it might seem a self-evident truth that 1 + 1 = 2, but a mathematician would have to prove that this is the case before it could be accepted.

Axioms are even more basic.

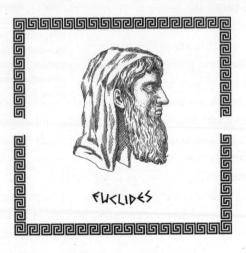

EUCLIDES

The Greek mathematician Euclid of Alexandria, working around 300BC, stated five axioms (or 'postulates') in the book *Elements*, usually attributed to him. (*Elements*, incidentally, turned out to be the most enduring book ever written that is not a religious text; it has been used to teach geometry for more than 2,000 years.)

1. Given any two points, you can draw a straight line between them. (This makes a 'line segment'.)

2. Any line segment can be extended indefinitely – meaning you can keep making a line longer without limit. (See, there really are things that are self-evidently true!)

3. Given a point and a line segment starting at the point, you can draw a circle centred on the point with the line segment as its radius. (This sounds harder, until you visualize it. The point is where you put the point of a pair of compasses. The line segment is the distance by which you extend the leg of

the compasses. Now you can sweep the compasses round, drawing a circle.)

4. All right-angles are equal to one another.

5. Given two straight lines, draw a line segment so that it crosses both. If the angles made with the lines, on the same side, add up to less than 180°, the two original lines eventually meet. This sound horribly complicated, but means that if you have a drawing like this:

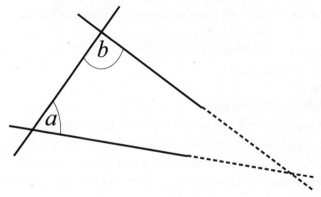

and a+b add up to less than 180°, then the lines are going to cross and make the shape into a triangle.

Euclid also set out five 'common notions':

- Things which are equal to the same thing are also equal to one another (that is, if a=b and b=c, then a=c)

- If equals are added to equals, the wholes are equal (that is, if a = b, then a + c = b + c)

- If equals are subtracted from equals, the remainders are equal (if a = b, then a - c = b - c)

- Things which coincide with one another are equal to one another.

- The whole is greater than the part.

Euclid was particularly concerned with geometry and his postulates are geared towards that use. In recent times, mathematicians have tried to make axioms as content- and context-free as possible.

The less relationship mathematical statements have to any specific situation, the more useful they are generally. For the average non-mathematician in the street, though, the less useful they appear the further they get from anything that looks like a real-world application.

Putting it to the test

How does proof work? Let's look at a very familiar theorem, Pythagoras's theorem. It states that if you square the lengths of each of the sides of a right-angled triangle, the squares of the shorter two will add up to the square of the longer side. (This is usually expressed as 'the sum of the square on the hypotenuse in a right-angled triangle is equal to the sum of the squares on the other two sides.')

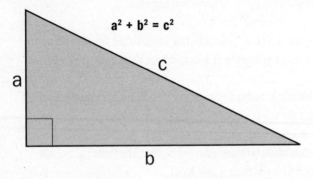

$$a^2 + b^2 = c^2$$

How can we prove this theorem? There are quite a few ways, but just one will do for now.

First, we'll draw a square using four triangles, like the one in grey on the right.

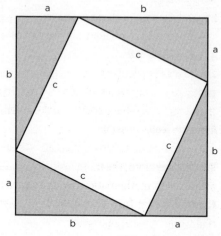

The right-angles of the triangles become the corners of the square. So now we have a big square with a smaller one inside it. You can possibly already see one proof immediately, just by looking at the illustration. Each side of the big square is given by a + b, so the whole area of the big square is:

A = (a + b)(a + b)

The area of each little triangle is:

½ x ab

The area of the square in the middle is:

c²

So we have two ways of writing the area of the whole square:

A = (a + b)(a + b)

and

$$A = c^2 + 4 \times (\tfrac{1}{2} \times ab)$$

Expanding both of these, we have:

$$A = a^2 + 2ab + b^2$$

and

$$A = c^2 + 2ab$$

We can therefore write:

$$A = a^2 + 2ab + b^2 = c^2 + 2ab$$

Taking the 2ab from each side:

$$a^2 + b^2 = c^2$$

Ta-dah! (or, more formally, QED).

It is because we can show that it's true, using the variables a, b, c to stand in for any number, that this counts as a proof and Pythagoras's theorem can be called a theorem. We don't need to try it out with every conceivable triangle, as the proof shows that it will indeed work with any right-angled triangle we can come up with, however large or small. The triangle could have sides of one nanometer and 40 billion kilometres and it would still be true.

So, difficult questions are hard to answer because intuition, 'it's obvious' or empirical evidence are not good enough to persuade a mathematician.

CHAPTER 6

What did the Babylonians ever do for us?

What time did you get up? At what angle were the hands of the clock? What's your star sign? Some of our everyday conventions are much older than you might think.

Start with 60

The Babylonian number system was organized around 10 and 60. Although it's often referred to as a base-60 system, it does also use 10 as a break-point (see Chapter 4). The Babylonians used only two symbols to represent numbers. They repeated the symbol for a unit (1) until they had nine of them, then used another symbol for 10. They used the units and 10s symbols in conjunction until they got to 60, then reused the 1 symbol in a different position. This meant that with just combinations of two symbols they could make any number by varying the position.

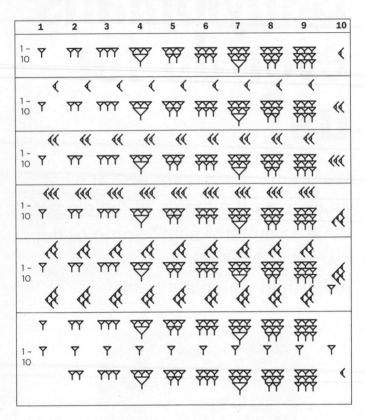

The place for 60s could be used up to 59, then another place was used for multiples of 3,600.

The spacing was crucial. The number 𝖸𝖸 is 2 x 1 = 2, but if there is a space between them, 𝖸 𝖸, the meaning is (60 x 1) + (1 x 1) = 61. There was a zero, represented by a slanting figure, but it could only be used to represent a zero in the middle of a number.

𝖸   **= 60 x 60 = 3,600**

𝖸 𝖸   **= 3,600 + 60 = 3,660**

𝖸 ◁ 𝖸   **= 3,600 + 0 + 1 = 3,601**

Seconds and minutes

The division of an hour into 60 minutes and a minute into 60 seconds comes from the Babylonian number system, though the Babylonians could not measure time that accurately. There are 360 degrees in a circle. Degrees are, in turn, divided into 60 minutes and those into 60 seconds. It would be very difficult to expunge 60 from our systems now, 4,000 years later. It is even finding its way into new systems that would be beyond the wildest imaginings of the Babylonians. The extent of the observable universe is measured in gigaparsecs (see Chapter 15). The definition of the parsec is based on the division of angles into 360°, and the subdivisions of 60 minutes and 60 seconds.

Why 60?

Sixty is a useful number for a base as it has many factors (2, 3, 4, 5, 6, 10, 12, 15, 20, 30). An important factor is 12 (60 = 12 x 5), and the Babylonians used this extensively, too. What the Babylonians (and before them the Sumerians) started, the Ancient Egyptians continued. They divided the day into twelve hours – twelve in the day and twelve in the night. The hours were different lengths at different seasons of the year, as the period when it was light was divided into twelve equal parts and the period when it was dark into another (often different) twelve equal parts.

It was the Ancient Greeks who first thought of having hours of equal length, but their ideas didn't really catch on until the Middle Ages and the advent of mechanical clocks. For the Babylonians, living fairly near the equator, hours were not of radically different lengths through the year. Perhaps if the Babylonians had lived in Finland they would have settled on equal hours from the start.

Minutes and seconds were introduced in AD1000 by the Arab polymath al-Biruni. The second was defined as $1/86,400^{\text{th}}$ of a mean solar day. It wasn't possible to measure time accurately at this point, though, and minutes and seconds were not relevant to most people for many more centuries.

Time and space

Minutes and seconds are used to measure both angles in geometry and intervals of time. Their use for angles came first,

and their connection with time came from the use of circular
time-keeping devices.

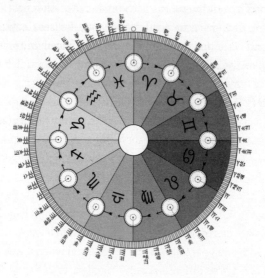

The Greek astronomer Eratosthenes (*c.*276–*c.*194BC) divided
a circle into 60 parts in an early version of latitude. This had
horizontal lines running through well-known locations (though
on a much smaller known world). Around 100 years later,
Hipparchus added a system of longitude lines that encompassed
360° and ran north to south, from pole to pole. About another 250
years later, around AD150, Ptolemy subdivided each of the 360°
into smaller segments. Each degree was divided into 60 parts,
each of which was again subdivided into 60 smaller parts. The
terms 'minute' and 'second' come from the Latin *partes minutae
primae*, or 'first very small part' and *partes minutae secundae* or
'second very small part'.

Half- and quarter- hours

In the 14th century, clock faces were divided into hours but not
minutes. Hours were divided into quarters and halves, which

is where the tradition of clocks striking at those intervals originates. Reliable measurement of minutes and their common inclusion on clock faces came at the end of the 17th century with the invention of the pendulum in 1690. Because a clock has a round face, and the hour was already divided into four, the decision to divide it into 60 minutes was entirely logical. It meant that each minute is 6°, and each second 0.1° – though it would take a very large clock face to be able to mark seconds legibly.

CHAPTER 7

Are some numbers too large?

Numbers are generally useful – but some are too large to serve any practical purpose.

As a small child, you might have set out to count to a million. If so, you probably gave up long before you got there.

How long would it take? If you counted one number a second and didn't stop to sleep, eat or rest in any way, it would take eleven-and-a-half days. That's not totally impossible. If you did the eating and sleeping and worked a bit less than half the time on your number task, it could be done in a month. If you succeeded, you might think of trying a billion. But that's not a good idea. At the same rate of one number a second, day and night, it would take thirty-one years and eight and a half months.

We never really grasp the difference between large numbers; it's easy to forget how quickly they escalate. If you think counting to a billion would be a boring way to spend 31 years, what about counting to a trillion? It would take more than 31,700 years. If you had started at the end of the last Ice Age, you wouldn't be a third of the way there yet – still only at around 300,000,000,000.

On the day I wrote this chapter, the US national debt was a bit over $18 trillion. Actually, that 'bit' was $171 billion, so not a small amount on its own. Let's pretend the debt began accumulating around 575,800 years ago at the rate of $1 a second and with an interest rate of zero. Modern humans hadn't evolved yet. Perhaps a glyptodon borrowed the first dollar.

DATE			DEBT
575,800 years ago	Glyptodon		$1
200,000 years ago	Modern humans		$11.86 trillion

15,000 years ago	Humans in America		$15 trillion
9,650 years ago	Mammoths extinct on mainland		$17.87 trillion
4,485 years ago	Pyramids started in Egypt		$18.03 trillion
AD450	Roman Empire ends		$18.12 trillion
1620	*Mayflower* sets sail		$18.158 trillion
1776	USA gains its independence		$18.163 trillion

That gives some idea of how many a trillion is, but trillions are very low in the table of large numbers.

Saving paper

Writing out long numbers – even billions and trillions, which economists and bankers do every day – uses up paper or screen real-estate quickly. Large numbers are not very easy to read either – you have to count the digits before you know what to call the first number. It's easy to see that the following number is 2 billion:

2,000,000,000

But could you say the following number out loud without stopping to count the digits?

234,168,017,329,112

Scientific notation makes writing large numbers easier. Instead of writing 1,000,000 for one million, we write 10^6, or ten to the power 6. This simply means 10 multiplied by itself 6 times:

10 x 10 x 10 x 10 x 10 x 10

10 x 10 = 100
100 x 10 = 1,000
1,000 x 10 = 10,000
10,000 x 10 = 100,000
100,000 x 10 = 1,000,000

So 10^6 is 1 followed by six zeros. A billion is 10^9 or 1 followed by nine zeros. And a trillion is 10^{12} – that's a lot easier to write and read than 1,000,000,000,000!

illions and [n]illions

A trillion is nowhere near the end of the 'illions'. We have:

Quadrillion	10^{15}
Quintillion	10^{18}
Sextillion	10^{21}
Septillion	10^{24}
Octillion	10^{27}
Nonillion	10^{30}
Decillion	10^{33}
Undecillion	10^{36}
Duodecillion	10^{39}
Tredecillion	10^{42}
Quattuordecillion	10^{45}
Quindecillion	10^{48}
Sexdecillion (Sedecillion)	10^{51}
Septendecillion	10^{54}
Octodecillion	10^{57}
Novemdecillion (Novendecillion)	10^{60}
Vigintillion	10^{63}
Centillion	10^{303}

Understanding the names

It might seem odd that a centillion has 303 zeros – shouldn't it have 100 zeros?

The Latin number-prefix (bi-, tri- and so on) does not show the number of zeros, it shows how many extra groups of three zeros there are in a number beyond the three zeros in a thousand.

So a million (1,000,000) has one group of three zeros more than a thousand.

A billion (1,000,000,000) has two extra groups of zeros, hence the prefix bi-.

A trillion has three extra groups of zeros.

A centillion has 100 extra groups of zeros, plus the original three in 1,000, giving a total of 303 zeros.

How high can you go?

There are two famous numbers that don't fall into the ...illion sequence: googol and googolplex. A googol is 1 followed by 100 zeros. At least we can write it:

10,000,000,000,000,000,000,000,000,000,000,000,000, 000,000,000,000,000,000,000,000,000,000,000,000,000, 000,000,000,000,000,000,000,000

A googolplex is an unimaginably large number: 10 to the power googol. It's written 10^{googol}. The terms 'googol' and 'googolplex' were invented by Milton Sirotta, the nine-year-old nephew of American mathematician Edward Kasner. He originally described googolplex as a 1 followed by as many zeros as you could write before you got fed up with the task.

A googolplex is so large that printing it would take longer than the entire history of the universe and use more than all the matter in the universe to produce the printout. At 10 point (the size of magazine print), the printout would also be 5×10^{68} times longer than the distance across the known universe.

To all intents and purposes, a googolplex might as well have stuck as 1 followed by the longest number of zeros you could write without getting tired, as it is an entirely useless number, at least in this universe.

Even a googol is more than is needed for any practical purpose. The number of elementary (that is, sub-atomic) particles in the universe is estimated at 10^{80} or 10^{81}. Since even one googol is 10,000,000,000,000,000,000 times that many – the number of sub-atomic particles in 10 quintillion universes like ours – a googolplex really is a bit much.

Some mathematicians have put their minds to working out how to represent numbers that are tiresome to write out even in scientific notation. If you get fed up with writing out the long,

long number of powers of ten you are using (for what?) you could try one of these methods.

The US mathematician David Knuth's notation uses ^ to indicate powers. The expression n^m means 'raise n to the power of m'. It is now commonly used on computers (in Excel, =10^6, for example, means 10^6).

n^2 = n^2	3^2 is 3^2 = 3 x 3 = 9
n^3 = n^3	3^3 is 3^3 = 3 x 3 x 3 = 27
n^4 = n^4	3^4 is 3^4 = 3 x 3 x 3 x 3 = 81

But Knuth allowed its use repeatedly. Doubling the ^ symbol to ^^ n means 'raise n to the power of m the following number of times'.

So, while

3^3 is 3^3 = 27

3^^3 is 3^(3^3) = 3^{27} = 7,625,597,484,987 – we've already got into the trillions!

And tripling the ^ symbol to ^^^ soon leads to very large numbers: **3^^^3 is written as 3^^4** and it is

3^3^3^3 = 3^3^{27} = $3^{7,625,597,484,987}$

The numbers rapidly get harder to read (as well as unimaginably large). People have come up with ways of writing even larger numbers – numbers you will never need to use. They don't even look like numbers, with single digits written inside different shapes, such as triangles and squares.

Now you can make it up

We can continue to make up larger and larger numbers. What about Graham's number squared (see box below)? Or 10 to the power of Graham's number? There is no end to the numbers we can name. Does that mean they exist, in any meaningful sense?

THE LARGEST NUMBER EVER

The largest number that has been used in any mathematical problem is called Graham's number. It's so large it's impossible to write it in any sensible way. It was suggested as the upper limit of a possible solution to a problem, but mathematicians think that the real answer to the problem might be '6'. It rather looks as though mathematics is getting back at us now and saying, 'Yeah, well. Whatever. Six will do.'

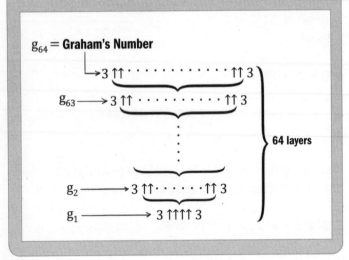

CHAPTER 8

What use is infinity?

If really, really big numbers are effectively useless, how much more useless is infinity?

At first it would seem that either the universe is infinite or it is limited. If it's limited, surely it can't contain anything infinite, can it? Well, it can. But first let's look at infinity in a bit more detail.

Endless numbers

If you ask most people what infinity is, they will think of the endless stream of numbers starting with 1, or perhaps 0, and going through 1,000,000, through googol and even googolplex, trudging ever onwards. We can always add another '1', always change a 1 to a 9, always multiply the number by itself – there is no end.

That's all true. But there's not just an infinity of numbers starting at 0 and going up, there is also an infinity of negative numbers – that is, numbers starting at 0 and going down.

How many infinities?

Just in case that wasn't enough, there is also an infinity of fractions (1 over a googol, etc.), and an infinity of decimal fractions (0.1, 0.11, etc.). No sooner have you got to 0.1111 recurring to infinity than you realize there is going to be 0.121111 to infinity, and so on, so there are lots of infinities even just between 0 and 1. There are just as many infinities between 1 and 2, and between 0 and -1. There is, inevitably, an infinity of infinities.

How big is infinity?

A question, familiar from curious children, is how big is infinity? This takes on a new dimension of complexity once we start to think about multiple (infinite) infinities. Common sense says the infinity of even numbers must be half the size of the infinity of all integers, and equal to the infinity of odd numbers. And yet they all go on forever. There is an infinity between each pair of numbers on the number line, and an infinity of digits in each irrational number. But surely the infinity between 1 and 2 can't be the same size as that between negative and positive infinity? The astonishing discovery that there are bigger and smaller infinities was demonstrated by Georg Cantor in 1874 and again in 1891.

Johannes Wallis, S.T.D.
Geometriæ Professor Savilianus Oxoniæ

Containable infinities

We tend to visualize infinity as extending into the void, so the idea that an infinity can be contained – between 0 and 1, for instance – is novel. Even so, if you visualize the infinity between 0 and 1, you might still imagine a line of numbers extending into the distance. The limit is never reached.

A more graspable infinity can be drawn from fractions, though.

A fractal is an infinitely replicating pattern, and it's a visible or visualizable infinity. A classic example of a fractal is the Koch snowflake. Begin by drawing an equilateral triangle (one with three equal sides). Then, in the middle third of each side of the triangle, draw another equilateral triangle, using that section as the base. Rub out the base, giving a star (hexagram in mathematicspeak). Do the same to each smaller triangle. And so on, and on (see page 76).

Each time you draw a new set of spiky triangles, the perimeter of the shape increases by one third. (Think about it – you are rubbing out one third of a side and adding the same length twice; one side of the new spike cancels out the bit removed, and the other side is a brand new bit of perimeter, a third the length of one side.) It's clear that the perimeter will carry on getting larger and larger, for although each section added is smaller and smaller, there are more and more of them.

If the length of an original side is s and the number of iterations is n, the total perimeter (P) is given by the expression:

P = 3s x (4/3)^n

As n increases, the perimeter tends towards infinity (because 4/3 is larger than 1 and so (4/3)^n keeps getting larger).

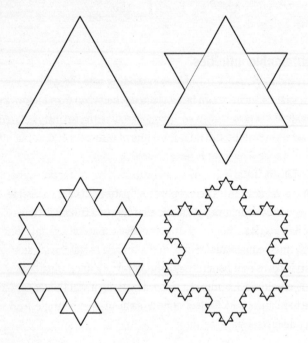

The area enclosed by each new triangle increases by one ninth of the increase added by the previous new triangle. This means that if the first triangle had area 9 cm², each point of the star will have area 9 ÷ 9 = 1 cm², and there are three new points, so the whole star has area 9 + 3 = 12 cm². With the first snowflake shape, each new triangle adds 1 ÷ 9 = $\frac{1}{9}$ cm² and there are 12, so the whole snowflake will have area

12 + (12 x $\frac{1}{9}$) = 12 + 1 $\frac{3}{9}$ = 13 $\frac{1}{3}$

SPOT THE FORMULA

The formula for an initial triangle with area a_0 is (close your eyes if you don't like formulae):

$$A_n = a_0 \left(1 + \frac{3}{5}\left(1 - \left(\frac{4}{9}\right)^n\right)\right)$$
$$= a_0 \left(8 - 3\left(\frac{4}{9}\right)^n\right)$$

As $\frac{4}{9}$ is less than 1, $\left(\frac{4}{9}\right)^n$ keeps getting smaller, so the area reaches a finite limit. In fact, it tends towards $\frac{8}{5}$ of the area of the original triangle.

Fractal action

There are many other designs of fractal, one of the most famous being the Mandelbrot pattern, derived from a complex series of numbers.

Fractals or near-fractals are also common in nature, occurring in structures that benefit from maximum surface area for a limited volume. Examples are the structure of blood vessels or tree roots, the branching of alveoli in the lungs, the structures of river deltas, mountains and even lightning.

A computer-generated image derived from a Mandelbrot set shows the elaborate and infinitely complicated boundary characteristic of this fractal shape, one of the best known examples of mathematical beauty.

Limited infinities

Although these patterns could theoretically repeat to infinity, they don't, of course, do so in nature. At some point we would reach the limiting size of molecules and be unable to repeat a pattern. They describe a process or series that could be infinitely extended but – as far as we know – there is nothing that is actually infinite. Even so, the infinite, and the infinitesimally small, can be a useful idea in mathematics, as we will see in Chapter 26.

CHAPTER 9

Are statistics lies, damned lies or worse?

We should be able to trust statistics, but the way they are presented is often designed to manipulate.

The media are full of statistics, many compiled in a way designed to persuade us to adopt a certain point of view. It's possible to avoid being manipulated if you understand not only what the statistics actually mean, but how we respond to numbers. There is as much psychology as mathematics involved.

Ways of looking at statistics

There are lots of ways of saying the same things about numbers, and they can elicit different responses from us. Journalists, advertisers and politicians can nudge us towards a particular interpretation depending on how they present figures.

All of the following mean the same thing:

- **1 in 5**
- **0.2 probability**
- **20% chance**
- **2 out 10**
- **5:1 odds**
- **10 in 50**
- **20 in every 100**
- **200,000 in a million**

Even so, we tend to respond differently to them. The last, 200,000 in a million, immediately sounds more impressive because the first number we read is large. Indeed, 20 out of 100 sounds more impressive than 2 in 10 because we think of 2 as a small number. This is a well-documented finding, called *ratio bias*. It can even lead people to choose a smaller probability of winning.

The following experiment demonstrates ratio bias well. People are presented with two bowls of glass beads made up as follows:

- **a bowl of ten beads, of which 9 are white and 1 is red**

- **a bowl of 100 beads, of which 92 are white and 8 are red**

The people are told they should hope to pick a red bead, but that they will be blindfolded. Which bowl should they choose, to increase their chance of picking a red bead?

In this test, 53 per cent of people choose the bowl with 100 beads in.

It's the wrong choice: the probability of picking a red bead in the first bowl is 10 per cent (10 in 100, or 1 in 10) but in the second bowl the probability is only 8 per cent (8 in 100).

The fact that there are more red beads in the second bowl seems to suggest that this means there are more opportunities to pick a red bead, and this appeals to some people. They completely ignore the fact that there are also more – disproportionately more – opportunities to pick a white bead. The chances of picking a red bead from the bowl of 100 are lower than the chances of picking a red bead from the other bowl. It seems that half the people tested did not understand how to choose to maximise their chances of picking a red bead.

High numbers work harder

People consider large numbers to be more significant than smaller numbers.

A sample of people asked to rate their perception of the severity of cancer as a health risk were split into two groups. Those who were told that 36,500 people a year die of cancer saw it as a more significant risk than those who were told that 100 people a day die of cancer.

In another study, subjects were more alarmed when told that 1,286 out of 10,000 people will die of cancer than when they were told that cancer will kill 24 out of 100 people, even though the second risk is nearly double the first one (24 per cent against 12.9 per cent).

This bias can lead people to make dangerous choices. When asked whether they would take a treatment option with a known risk of death, people's answers depended on how the figures were presented.

If the number of deaths of previous patients was shown as a rate per 100 patients, the subjects tolerated a much higher risk than when they were presented as deaths per 1,000. Potential patients would accept a risk of death of up to 37.1 per cent in the first case, but only up to 17.6 per cent in the second.

The larger number (176 against 37) blinded them to the lower *level* of risk.

Don't look underneath!

If asked which of several fractional numbers is larger, people tend to compare only the *numerator* (the number on top of the fraction) and ignore the *denominator* (the number below). This is why people prefer a probability of 8/100 than of 1/10 when picking beads. Completely disregarding the overall number in this way is called *denominator neglect*.

If you are commercially minded, you can use this to your advantage. Imagine you are running a fete to raise money for a charity and want to persuade people to pay for a chance of winning in a game. You can exploit denominator neglect or ratio bias to encourage people to play in games that have a *lower* chance of success but look as though they have a higher chance of success. Instead of saying '1 in 10 wins a prize', you will get more takers with '8 in every 100 wins a prize!' (Adding the '!' is not mathematical, but it does help because it is a cue to the reader to be surprised or impressed.)

What are they not saying?

Another way in which politicians, advertisers and journalists manipulate the way we think about figures is by careful choice and phraseology. Try inverting every sentence that uses figures and see what it really means:

- **30 per cent of people are worse off under this government = 70 per cent have at least as high a standard of living under this government as under the last.**

- **1 in 4 laptops breaks within 24 months = 3 in 4 laptops are still working after 24 months.**

- **30 out of every 50 residents live to be over 70 = 40 per cent of residents die before reaching 70.**

By choosing which half of a mathematical statement to concentrate on, the presenter can encourage us towards a positive or negative view. They can reinforce this effect by choosing a presentation method that makes it harder for us to see the other side of the story. If the last example – 30 out of every 50 residents live to be over 70 – had been stated as '60 per cent of residents live to be over 70', we could see at a glance that this leaves 40 per cent who die before this age. But 30 is a largeish number, and we have to do the mathematics (50 – 30, then convert 20 to a percentage) to see the real state of affairs.

Look for the context

Another trick is to give one statistic on its own. Numbers out of context are pretty meaningless. If you read that 20 pupils in a school have been suspended for drug misuse, that sounds pretty bad. But it is far worse if the school has 800 pupils than if it has 2,000 pupils. If 20 pupils in a school of 2,000 misuse drugs, that means 99 per cent of the pupils don't misuse drugs. That's not much of a headline, though.

'The odds are a million to one that . . . ' is a fairly common way for the media to say that something is very unlikely. Strictly speaking, it is unlikely in any particular instance, but if there are lots of instances it is not unlikely overall. If the chances of an African elephant being born albino are one in a million, we would be unlikely ever to see one if we visited Africa. If the chances of an ant being albino are one in a million, it would be quite surprising if we didn't see at least one if we turned over a few anthills.

Apples and oranges

It's difficult to compare statistics at a glance if numbers are presented in different ways. Media reports often do this – possibly to confuse us, but possibly just because the journalist thinks it looks a bit more varied. Comparing information from different sources often causes this problem, but that's still sloppy – the journalist should make them comparable. For example, it's hard to understand a news report which says 2 out of 10 people do enough exercise to reduce their risk of heart disease by 30 per cent and a further third of people do enough exercise to cut their risk by 15 per cent. It asks us to think about figures in three different ways: 2 out of 10, fractions and percentages. The data would be much clearer if the figures were all converted to percentages: 20 per cent of people do enough exercise to reduce their risk by 30 per cent and a further 33 per cent of people cut their risk by 15 per cent. This also makes it easy to see that 47 per cent of people don't do enough exercise:

100 – (20 + 33) = 100 – 53 = 47

CHAPTER 10

Is that significant?

Do the facts and figures really show
what they claim to show?

Statistics have an air of authority and people are easily swayed by them. They look like 'proof', even when they do not actually prove anything.

Significant or not?

Statisticians need to know whether the facts and figures generated by research, studies, surveys or whatever are 'significant'. In other words, do they provide useful information that people can act on, or could that result have occurred by chance or through errors in choosing the sample? In general, scientific studies are considered to have found a significant result if the probability (p) that a result might be random or erroneous is less than 1 in 20. This is expressed as:

$p < 0.05$

when p stands for probability. A probability of 1 means something is absolutely certain: there is a probability of 1 that if you are reading this book you are alive. Probability of 0 means something is definitely not happening. There is a probability of 0 that your copy of the book is printed on water.

The probability $p < 0.05$ is defined in a rather odd way. It is a 5 per cent chance that the 'null hypothesis is true', and the null hypothesis is that there is no effect. Wading through the double negatives, it means that as long as the chance that the results are a coincidence is less than 5 per cent, the statistics are good. The 5 per cent margin is often also used to disregard outliers – samples that fall outside the main body of results.

The curve in the picture below shows the usual – or normal – pattern of distribution of results (there is more about it in Chapter 14). The results that are generally considered valid and so can be included in further processing are those that fall in the middle 95 per cent. In some studies, more precise or rigorous tests of significance are required. This is applied to really important studies – those which will redefine science. The probability required for the detection of the Higgs Boson (a type of subatomic particle) to be confirmed, for example, is set at about 1 in 3.5 million, or $p < 2.86 \times 10^{-7}$.

ALL SWANS ARE WHITE – OR ARE THEY?

Long ago, Europeans thought all swans were white because they had never seen a black swan. The sample size was huge – pretty much all the swans in Europe. But you only need to see one black swan to scupper that theory. The Viennese-born British philosopher Karl Popper (1902–94) developed a definition of science that requires theories to be falsifiable – that is, capable of being shown to be wrong – in order to count as scientific. The theory that all swans are white is indeed falsifiable (by seeing a non-white swan), so it can be proposed as a theory. But it is not verifiable. We can't prove it is correct without observing all the swans in the world through all time. This is why you can't prove a negative. Just because you haven't seen something doesn't mean it doesn't exist. For this reason, the opposite of the idea proposed – the null hypothesis in these statistical examples – is an important test.

No effect? Or not significant?

If a study finds there is no 'statistically significant' result, that doesn't necessarily mean there is no effect. It's important also to look at the size of the sample and the design of the study.

A small-scale study might not pick up a small effect. The timescale could be too short, or the sample size too small. This is something drug trials have to take into account, for instance. A study involving only 20 subjects will not be able to show something that affects only 2 per cent of the subjects – either it will appear to affect none, or it will appear to affect 1 (or more) in 20, so 5 per cent or more.

Correlation and causality

News stories often make links between behaviour and events, suggesting that one causes the other. We might read that people who wear a cycle helmet are less likely to suffer serious head injury in a cycling accident, for instance. The suggestion is that the cycle helmet protects them, and that is very likely true. But it's also possible to present two sets of figures to suggest a connection that might not exist, or that might be different from the implied connection. For example, newspaper purchases and the murder rate have both fallen over the last five years. There is a correlation here – the patterns are similar. However, presenting the figures side by side might suggest that the two are related – does buying a newspaper send people into a murderous rage? Probably not. There is a correlation but no causality: one does not cause the other.

In the winter, sales of sledges increase and sales of ice cream decrease. Here there is a link, but it's not direct: both are related to the weather, but not to each other. Beware of graphs and tables of statistics that seem to suggest a link between two phenomena – there might be a link, but there might also be other factors at play, known as confounding variables, which link with both. In the example of sledges and ice cream, the weather is the confounding variable. There is not always a confounding variable – in some cases it could just be coincidence.

PROBABLY NOT . . .

There are correlations between:

- sales of organic food and diagnoses of autism
- the use of Facebook and the Greek debt crisis
- the import of lemons from Mexico and the rate of deaths on American roads – this is an inverse correlation: deaths go down as lemon imports rise
- the decline in the number of pirates and increase in global warming – this is also an inverse correlation: did pirates prevent global warming?

www.buzzfeed.com/kjh2110/the-10-most-bizarre-correlations

CHAPTER 11

How big is a planet?

What if you suddenly found yourself
stranded on a different planet?
Could you work out how big it was?

That might not be your first concern, of course, but just suppose for a moment.... How do you measure the size of something that is too large to pace out?

Round or flat?

Contrary to popular legend, very few people have ever thought the Earth is flat. The simple fact that you can see something appearing over the horizon shows that it can't be flat. Someone standing on a seashore and watching a ship approaching can see that the mast – the tallest part – appears first, and then the rest of the ship slowly appears over the horizon. This can only happen if the surface of the Earth is curved. If the Earth were flat, a distant object would still appear tiny, but its full height would be visible immediately and only its size would increase as it approached.

You don't even need to be near the sea – which is just as well, as that alien planet might not have a sea or any ships. The fact that you can see further from a high vantage point than from low down also shows that the surface of the Earth is curved.

Where is the horizon?

If you stand on a flat plain, or at sea level looking across the sea, the furthest you can see at the same level (on Earth) is 3.2km 2 miles).

This assumes your eyes are at 'eye level' (that is, you're not lying on the floor) and that you are about 1.8m (5ft 10in) tall. You can see the top of taller objects that are further away. If you stand on a hill, or on the deck of ship, you can see further than 3.2km (2 miles).

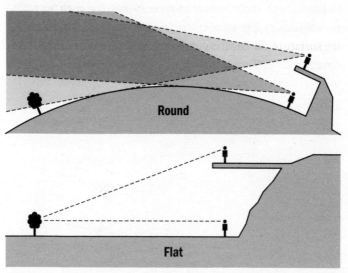

Round

Flat

All the way round

The size of the Earth preoccupied people long before they had good technologies for measuring it. The Ancient Greek philosopher Eratosthenes is the first person we know of who tried to calculate the circumference of the Earth. He lived in Alexandria, Egypt, and carried out his calculation around 240BC.

Eratosthenes knew that in the nearby city of Syene there was a well which, if he peered into it at noon on the summer solstice, had no shadow at the bottom. If there was no shadow, that must mean that the sun was directly overhead and so could shine straight down into the well. He also knew that there was no such shadow-free time in the wells in his own city at noon on that day. (This is because Alexandria is further north than Syene.)

Eratosthenes realized that if he compared the shadow in Alexandria with the lack-of-shadow in Syene, he could work out the circumference of the Earth. He measured the angle between a tall tower in Alexandria and the edge of its shadow at noon (when he knew there would be no shadow in Syene). The angle was 7.2°. He knew that when a line crosses two parallel lines, the inside angle on each side is the same. The rays of the sun are, to all intents and purposes, parallel as they originate so far away.

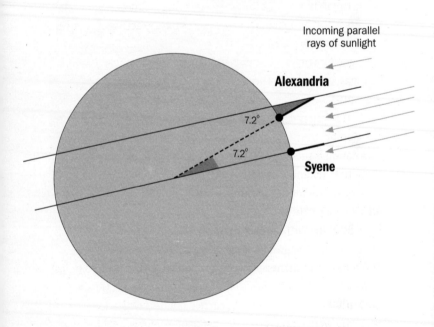

This meant that the angle at the centre of the Earth (which he assumed to be spherical) between lines drawn from Syene and Alexandria would be the same as the angle of the shadow cast by the tower. The ratio

full circle : angle measured

would be the same as the ratio

Earth's circumference : distance between Syene and Alexandria

Eratosthenes knew the distance between the two towns. Unfortunately, we don't know exactly what he knew it to be – he reports it as 5,000 'stadia', but we don't know exactly how long a 'stadion' was.

By good fortune, 7.2° represents $^1/_{50}$th of a circle (360 ÷ 7.2 = 50). This gave a circumference of 5,000 x 50 = 250,000 stadia.

Eratosthenes might have been right to within about 1 per cent of the true circumference, or, if he was using a different measure of a stadion, he might have been out by 16 per cent. Even so, his calculation is pretty good. Using his calculation of the angle and the actual distance between the cities, 800km (498 miles), the answer we get for the circumference is

50 x 800km = 40,000km

The actual circumference of the Earth is 40,075km (25,000 miles).

Stranded

So, if you were stranded on another planet, you would have two ways of discovering its size. To use Eratosthenes's method, you would need to find a place where the sun cast no shadow at noon and somewhere a manageable distance away where it did cast a shadow at noon; then you would need to measure the angle of the shadow, as he did. Of course you might not have taken a protractor, so this method could be tricky. An alternative method would be to measure the distance to the horizon.

To use the distance-to-the horizon method, you would need to measure, perhaps by pacing, how far you could walk from an object before it disappears over the horizon. There is an equation to help you work out how far you can see at different heights:

$$d^2 = (r + h)^2 - r^2$$

where d is the distance you can see, r is the radius of the earth, and h is the distance from your eyes to the ground (all distances in the same units).

This uses Pythagoras's theorem, which states that the square on the hypotenuse of a right-angled triangle is equal to the square on the other two sides added together (see page 56).

You can use this to work out the value of r (radius of the planet).

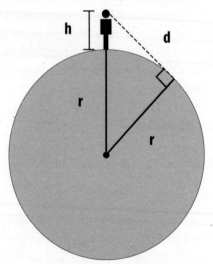

Expanding

$d^2 =$
$(r + h)^2 - r^2 =$
$r^2 + 2hr + h^2 - r^2 =$
$2hr + h^2$

So if you could see something 10km (6.4 miles) away and your own eye height was 1.5m (that is, 0.0015km),

$10^2 = 2 \times 0.0015r + 1.5^2$
$100 = 0.003r + 2.25$
$100 - 2.25 = 0.003r$
$97.75 = 0.003r$
$3,258 = r$

After that, you'd need to work out the circumference, $2\pi r$:

$2 \times \pi \times 3,258 = 20,473km$ (12,721 miles)

Don't set out to walk around that planet!

CHAPTER 12

How straight is a line?

The shortest route from A to B is certainly a line – but is it a straight one?

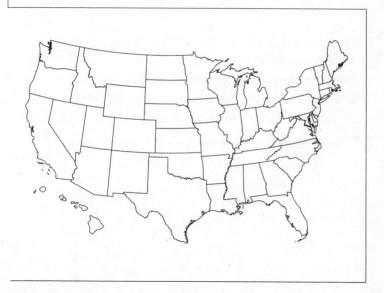

It's pretty obvious that on a flat plane, the shortest path between two points is a straight line. It is possible to prove it mathematically, but the proof, using differential calculus (see Chapter 26), is a bit too long and onerous for this book.

Lines, long and short

Imagine you are at A and want to get to B. The path might be wiggly, especially if you are following a map.

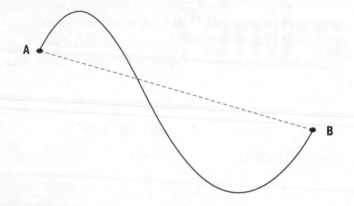

To make the wiggly path shorter, we flatten the curves. The flattest the curves can get is a straight line.

We can do it without curves, too. Any straight line can be made into the hypotenuse of a right-angled triangle – indeed of an infinite number of right-angled triangles.

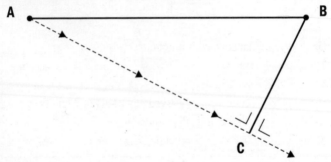

Whichever triangle you draw, the total AC + CB is always going to be larger than AB.

So far so, so good. But we don't live on a flat world.

On the ball

Euclid (see page 54) set out all the basics of geometry for a planar (flat) world. Euclidean geometry, as it's called, has lots of good, practical uses, such as working out the volume of skip you need in order to cart away soil when you dig a pond, or how many metres of carpet you need for a room. However, we live on a near-spherical Earth, where a straight line is not all it seems. Now we need to use some non-Euclidean geometry.

Euclid's fifth postulate (see page 55) demonstrates that parallel lines never meet, by showing the characteristics of lines that do meet. A line intersecting two lines is perpendicular (at right-angles) to both if the two lines are parallel:

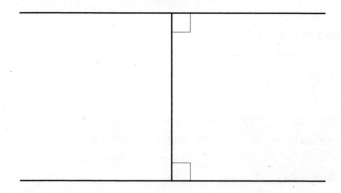

That's true on a flat surface, but not on a curved surface.

There are two types of curved surface – a concave surface, like the inside of a bowl, and a convex surface, like the outside of a globe. That gives two types of curved-surface geometry. They are called hyperbolic and elliptical geometries.

Now we can draw a line perpendicular to two other lines

without those other lines being parallel. On a hyperbolic surface, the lines curve away from each other in both directions, the distance between them increasing. On an elliptical surface, they curve towards each other and will eventually meet on both sides.

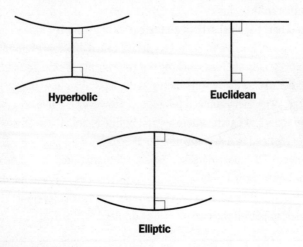

Hyperbolic

Euclidean

Elliptic

As the crow flies

We are used to thinking of geographical distances as being shortest 'as the crow flies'. That might be drawn on a map as a straight line.

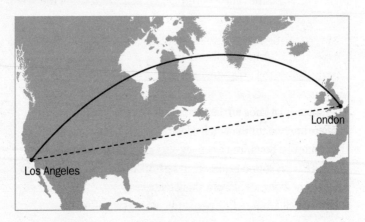

Here the (energetic) crow flying from Los Angeles to London might plan its route by looking at a map and drawing a straight line between the two cities. But if it took this route, it would actually travel further than if it followed the curved path, even though the latter looks longer. The reason becomes clear if we remember that the Earth is a spheroid.

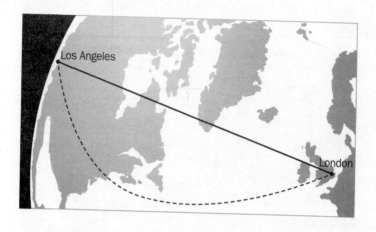

The shortest line between two points on a sphere follows a *geodesic*. A geodesic is a line that runs all around the sphere in a circle centred on the centre of the sphere. This means that the diameter of the circle is the same as the diameter of the sphere. A geodesic is also called a 'great circle'. We can draw any number of great circles around a sphere.

Getting back to Earth – all lines of longitude are great circles. No lines of latitude, except the equator, are great circles. All the other lines of latitude are lesser (or small) circles, with a radius smaller than that of the whole globe.

The shortest distance between two points on the surface of a globe is always found by drawing a great circle between the two points – the distance along a small circle is always longer (even if it doesn't look it).

The crow's map

The line that looks shortest drawn on a flat map is a small circle when the route is checked on a globe. The reason that the actual flight path of a crow or plane looks longer on a map than the apparently 'straight' path is that all map projections distort the geography of the world. It is not possible to draw the surface of a sphere on a flat plane without one kind of distortion or another. The most familiar one is the Mercator projection (below).

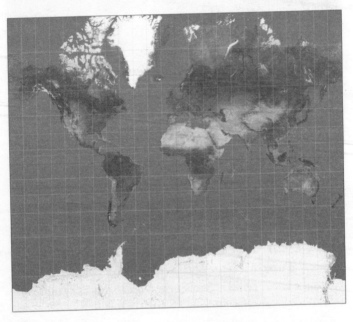

This is increasingly distorted the closer we get to the poles. One result is that Greenland looks much larger than it actually is, and Antarctica looks about the same size as all the warmer lands put together – it's actually less than one-and-a-half times the size of Australia.

In the Gall-Peters projection on page 105, which shows equal areas, the picture is very different. Now Greenland is really quite small, and Africa is far bigger. This projection has not been popular in North America as it makes this continent look a lot less important compared to South America, Africa and Australia than Americans are used to. Africa has three times the land area of the USA.

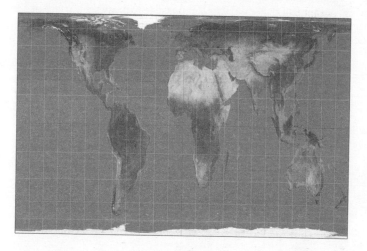

The distortion of the projections used for flat maps, combined with the translation of the great circle to a flat line, makes a direct flight path look like a circuitous parabola.

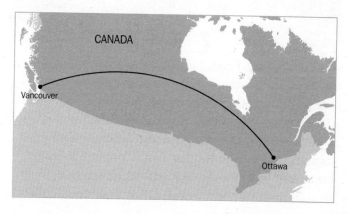

HOW BIG IS GREENLAND?

On the familiar Mercator map, Greenland looks about the size of Africa, and Antarctica looks larger than all the warmer countries put together. In fact, Greenland is about a fourteenth the size of Africa.

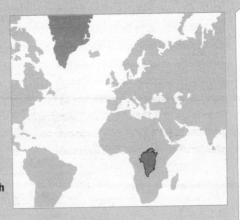

And Russia, which looks vast using Mercator's projection, is also smaller than Africa in reality.

Shorter isn't always faster or better. Planes don't always follow the most direct great-circle route because wind and air traffic patterns will also affect their choice of route.

We live in a real world, not the uncluttered paradise of mathematics, and there are always other factors to take into account, such as gravity, weather, air traffic control and even hostile forces on the ground with anti-aircraft weapons.

Adding complicating factors does not defeat mathematics, but it does make it more challenging. This is a puzzle set by Johann Bernoulli in the 17th century. Imagine a piece of wire with a bead threaded on to it. Which shape should you bend the wire into for the bead to fall most quickly from the start to the finish? (The wire is the same length in each case.)

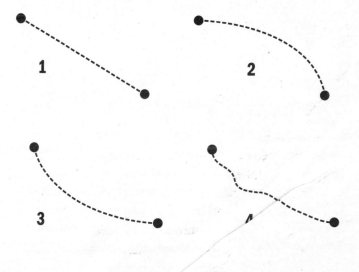

Many brilliant mathematicians, including Newton, Bernoulli, Huygens, and Leibniz tried this problem. Galileo got it wrong. The first one to give the correct answer was Newton, who had the advantage of having developed calculus.

The correct answer is the third shape: the steep downwards curve allows the bead to build up speed to cover the horizontal distance more quickly. A bead following this trajectory could travel further in the same time than a bead on a shorter, straight wire. So while the shortest distance might be a straight line on a flat plane, the quickest route might not be a straight line at all.

CHAPTER 13

Do you like the wallpaper?

If you were to look at a wallpaper catalogue, you could be forgiven for thinking there is a vast range of patterns available.

For mathematicians, however, there are only 17 basic patterns in the so-called 'wallpaper group'.

Let's see that again. And again

Mathematicians are not, actually, all that concerned with wallpaper *per se* – but they are interested in isometry (see below), and that's what lies behind the wallpaper group of patterns. The proof that there are only 17 basic patterns that comprise the wallpaper group was demonstrated by the Russian mathematician, geologist and crystallographer Evgraf Fedorov in 1891. All the patterns built up from repeating isometries are based on a 'cell' which must be a specific shape, usually rectangular (sometimes specifically square) or hexagonal.

A bit about isometry

You don't want the shapes on your wallpaper distorting, growing or shrinking as they move across the wall. That would give you nightmares. Instead, the copies of the pattern, even if they are turned round or reflected, must look the same. Mathematically, this is called isometry: the distance between any two points on the image must remain the same after the image has been transformed (that is, changed). It's easier to understand with an example. Here is a seahorse.

Here are some ways of changing the image of the seahorse:

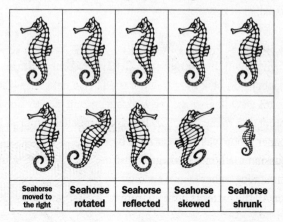

Seahorse moved to the right	Seahorse rotated	Seahorse reflected	Seahorse skewed	Seahorse shrunk

The first three are isometric transformations – absolute distances between any two points on the seahorse are the same before and after transformation. The fourth and fifth are non-isometric: skewing and shrinking change the distances between points.

There are four types of isometry in two dimensions:

translation (moving the figure wholesale to the left, right, up or down)

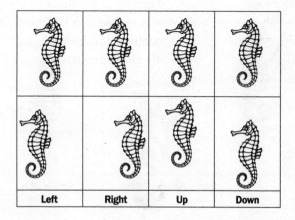

Left	Right	Up	Down

rotation (rotating the figure clockwise or anticlockwise)

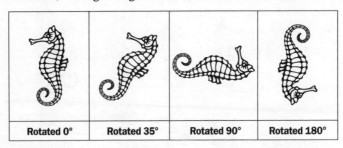

| Rotated 0° | Rotated 35° | Rotated 90° | Rotated 180° |

reflection (reflecting – making a mirror-image – of the figure in any direction)

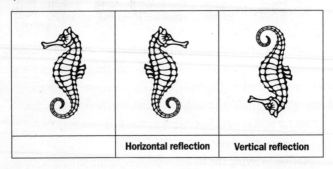

| | Horizontal reflection | Vertical reflection |

glide reflection (a combination of reflection and translation at the same time)

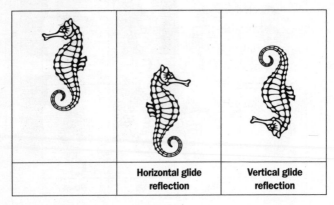

| | Horizontal glide reflection | Vertical glide reflection |

Mathematicians give the 17 patterns odd-looking names, nowhere near as appealing as the names in a wallpaper catalogue. The names are made of a code that explains how to make the pattern.

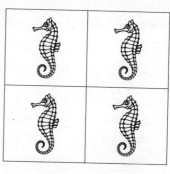

- p1 (above right) is the simplest form, with the image just translated in one direction. The cell shape can be any parallelogram (including a rectangle or square).

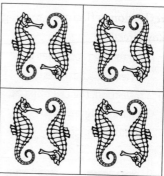

- p2 (below right) is similar to p1, but the tile can be turned upside down without altering the image

- pm (bottom right) can be reflected along one axis; this means the figure is symmetrical along an axis. The cell must be rectangular or square:

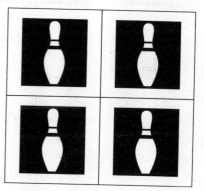

- pg (right) has a glide reflection – reflected and moved at the same time

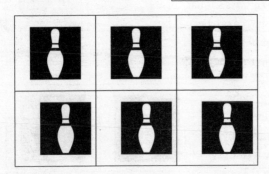

- cm (below) combines glide reflection and an axis of reflection; the cell must be a parallelogram with equal sides.

The tilings become increasingly complex as they combine reflections, rotations and glides in different directions. Interestingly, examples of all of them can be found in ancient art, including paintings on Egyptian mummy cases, Arab tiling and mosaics, Assyrian bronze work, Turkish pottery, Tahitian weaving, and Chinese and Persian porcelain. A few examples are shown on the facing page.

p2mg – cloth, Hawaii

p4 – ceiling of Egyptian tomb

p4mg – Chinese porcelain

p3m1 – Persian glazed tile

p31m – painted porcelain
from China

p6mm – bronze vessel from
Nimroud, Assyria

How about a frieze around that wallpaper?

The wallpaper group repeats the pattern in two directions – along the wall, as it were, and up and down the wall, from floor to ceiling. Another group, known as the frieze group, is repeated in only one direction, so it could be used to make a frieze to go along a wall.

The seven types, again, are all found in early art – and even prehistoric decorations:

p1	Horizontal translation	
p1m1	Translation, reflects vertically	
p11m	Translation, reflects vertically and horizontally	
p11g	Translation and glide reflection	
p2	Translation and 180° rotation	
p2mg	Translation, 180° rotation, vertical reflection and glide reflection	
p2mm	Translation, 180° rotation, horizontal and vertical reflection, and glide reflection	

And now the tiles . . .

The wallpaper group works with cells of a shape that can be tessellated – that is, repeated to cover a flat space leaving no gaps. Tessellation is another way of building patterns, working with the shape of the cells rather than a pattern drawn on them.

Again, most common tessellations are found in ancient art. The simplest tessellations use a single shape repeated. These are called regular tessellations.

The three basic types of tessellation are shown at the top of the facing page.

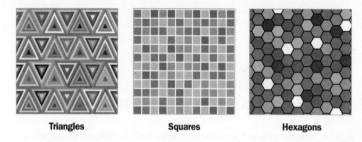

Triangles   **Squares**   **Hexagons**

The pattern is identical at each vertex (corner).

Tessellations are described by listing the number of sides of each of the shapes meeting at a vertex.

Each corner in the hexagonal pattern is shared by three hexagons. The hexagons each have six sides, so the tessellation is 6.6.6.

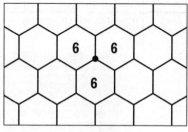

Semi-regular tessellations can have two or more shapes tessellated together. There are eight semi-regular tessellations (see below and at the top of page 118).

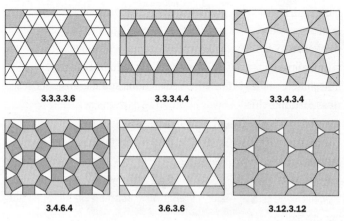

3.3.3.3.6   **3.3.3.4.4**   **3.3.4.3.4**

3.4.6.4   **3.6.3.6**   **3.12.3.12**

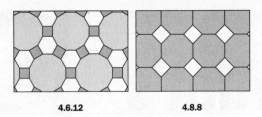

4.6.12   4.8.8

Again, the pattern at each vertex is the same, though it might be rotated.

Irregular tessellations don't have the same pattern at each vertex, so can't be described using the same system. They must still cover the whole surface without gaps or overlaps. This irregular tessellation is in the Alhambra palace, Spain:

You could use any of these tessellation patterns to tile your bathroom if you were proficient enough. More ambitious and artistic tessellations, often using curved shapes, have been developed by the Dutch artist M.C. Escher (1898–1972). The surface is still completely filled, but now with imaginative and sometimes nightmarish permutations and metamorphoses of shapes, like this:

CHAPTER 14

What's normal?

How heavy is a baby? How long is a boa constrictor? How often do people go to the supermarket?

The answer to the questions on the previous page is: 'It varies'. But although the answers differ from one baby, boa constrictor or supermarket shopper to another, there are boundaries within which we can expect any individual example to fall.

Human babies are not going to weigh three nanograms or five tonnes. The boa constrictor is not going to be 40km (25 miles) long. People don't go to the supermarket once a minute or once a millennium.

The average baby

Before a baby is born, the parents will have expectations about its likely weight, derived from their knowledge of other babies. After the birth, the baby's actual weight is established and compared with others.

Advance knowledge about the average weight is useful to parents ('Should I buy very small baby clothes?') and to healthcare workers ('Is this baby at risk?'). After the event, knowledge about averages is especially useful to healthcare workers to help answer questions such as: 'Is this baby so far from "normal" that we should be worried?'

On the right is a table of weights of some newborn babies.

A table of baby weights is hard to process mentally, even if they are listed in size order. It's easier to grasp the weight of babies from an average.

Mean babies

There are three types of 'average' we can calculate:

Baby	Weight (kg/lb)
1	2.3kg/5.1lb
2	2.3kg/5.1lb
3	2.9kg/6.4lb
4	3.0kg/6.6lb
5	3.2kg/7.1lb
6	3.3kg/7.3lb
7	3.4kg/7.5lb
8	3.5kg/7.7lb
9	3.7kg/8.2lb
10	3.8kg/8.4lb

The mean. This is what most people think of as the average. Add up all the values and divide by the number of values:

2.3 + 2.3 + 2.9 + 3.0 + 3.2
+ 3.3 + 3.4 + 3.5 + 3.7 + 3.8
= 31.4
31.4 ÷ 10 = 3.14kg (6.9lb).

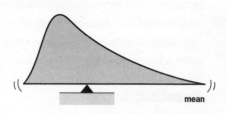

The median. This is the value in the middle of the range, meaning that half the values are above it and half below it. Arrange the values in order of size (as they are in the table) and pick the one in the middle of the list. If there is an even number of values there will be two values in the middle. The median is then the mean of these two, so the median here is the mean of 3.2kg (7.1lb) and 3.3kg (7.3lb), which is 3.25kg (7.15lb).

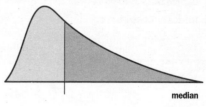

The mode. This is the value that occurs most frequently. There are two babies weighing 2.3kg (5.1lb), but only one example of all the other weights, so 2.3kg (5.1lb) is the mode.

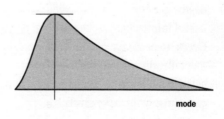

With a small dataset, like the table of imaginary babies on the facing page, the mode can be very misleading. Looking at these values, and going by the mode, we might expect a baby to weigh

2.3kg (5.1lb), but that is actually much lower than the weight of most newborn babies. As with all statistics, the larger the dataset, the more confidence we can put in any analysis of it. With a small dataset like this, the median and mean are more reliable and useful than the mode. Indeed, often there is no mode, as each value occurs only once.

Normal distribution

An easier way to look at a lot of data is in the form of a bell curve, below. At either end of the curve, a very small number of babies will be extremely small or extremely large, but the weight of most babies falls somewhere in the middle of the curve.

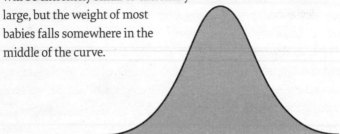

Deviating from the norm

But where on the curve are we going to call 'normal'? Clearly not only those examples that fall right in the middle. To be really useful, the curve needs to come with more information. The most useful information is the standard deviation, represented by the Greek letter sigma, σ. This is a measure of how far all the examples differ from the mean, expressed as the average, or standard. It is calculated using a formula that looks daunting but is fairly easy to use in practice:

$$\sigma = \sqrt{\frac{1}{N} \sum_{i=1}^{N} (x_i - \mu)^2}$$

It involves the following steps, starting inside the brackets:

- For each value, subtract the mean from each value: $x_i - \mu$
- square each difference: $(x_i - \mu)^2$

Then

- add the squared differences together: $\Sigma(x_i - \mu)^2$
- divide by the number of values you have – the result is called the variance: $1/N \, \Sigma(x_i - \mu)^2$
- find the square root of the variance: σ. That's the standard deviation.

The reason we square the values and then unsquare them at the end is that otherwise the negative values (instances where the value is lower than the mean) would cancel out the positive values. The standard deviation for our list of baby weights is 0.5kg (1.1lb).

WHICH FIRST?

When calculations involve several steps, it can be hard to know in what order to do the steps. The mnemonic BODMAS helps:

B – do anything inside brackets first

O – do anything to do with 'orders' next; this means raising numbers to powers or calculating square roots

D – do any divisions next, going left to right if there are more than one

M – do any multiplications next, again going left to right

A – do any additions next, left to right

S – finally, do any subtractions, left to right.

A sample or a population?

We've assumed that these babies are the whole population of babies under investigation. However, if we wanted to use this sample to find out about the weights of newborn babies in general, we would have to adjust the standard deviation calculation slightly. Instead of dividing by N, we divide by N-1. This makes the standard deviation slightly larger – 0.52kg (1.15lb) in our sample. It adds some flexibility because there will inevitably be more variation in a larger population than in a sample, unless by random chance you happen to have picked both the largest and the smallest examples from the whole population.

Looking at the table again, we find that only babies 1, 2, 9 and 10 lie more than one standard deviation (more than 0.52kg/1.15lb) from the mean of 3.14kg (6.9lb). So someone expecting a baby could reasonably anticipate that their baby might weigh between 2.6kg (5.7lb) and 3.7kg (8.2lb).

Baby	Weight (kg/lb)
1	2.3kg/5.1lb
2	2.3kg/5.1lb
3	2.9kg/6.4lb
4	3.0kg/6.6lb
5	3.2kg/7.1lb
6	3.3kg/7.3lb
7	3.4kg/7.5lb
8	3.5kg/7.7lb
9	3.7kg/8.2lb
10	3.8kg/8.4lb

Percentiles

With a study of a large population, we can find out more information.

Percentiles (or centiles) indicate what percentage of values fall below a particular level. The 50th percentile is the middle of the range, with 50 per cent of values being higher and 50 per cent lower. By the 90th percentile, 90 per cent of values are lower and only 10 per cent are higher. At the 2nd percentile, only 2 per cent of values are lower and 98 per cent of values are higher.

Percentile charts are often used to show the expected growth patterns of children.

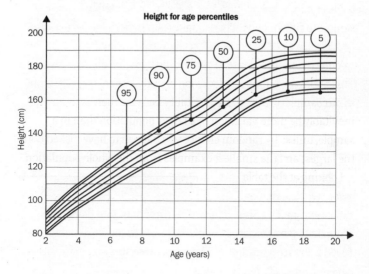

Height for age percentiles

A chart like this doesn't mean that there are never going to be children larger than the 95th percentile or smaller than the 5th percentile, but there will not be many: 90 per cent of all children will fit somewhere between the top and bottom lines on this chart.

Smaller or larger children might be watched, but there is not necessarily anything wrong with them.

A normal curve

Combining the idea of percentiles with the bell curve, we can chop the curve up into parts that are within one, two or three standard deviations of the mean. It just so happens that in many cases, the chopping-up looks like the graph on the next page.

This is called a normal distribution curve. Many things naturally fall into this pattern with 68 per cent of values within one standard deviation, 95 per cent within two standard

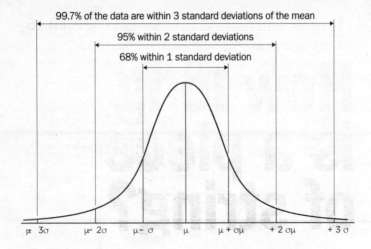

99.7% of the data are within 3 standard deviations of the mean

95% within 2 standard deviations

68% within 1 standard deviation

$\mu - 3\sigma$ $\mu - 2\sigma$ $\mu - \sigma$ μ $\mu + \sigma$ $\mu + 2\sigma$ $\mu + 3\sigma$

deviations and 99.7 per cent within three standard deviations. It applies to the heights of humans, errors in measurements, blood pressure readings, marks on exam scripts and many other sets of values.

Lying outside the band of two or three standard deviations from the mean can act as an alert. But the boundaries can also be used to set what is 'normal'. Suppose it's your job to set exams each year. You can't be certain that each year's exam paper is equally difficult or marked with equal rigour. What you can do, though, is set the pass mark according to a normal distribution curve. If you were to plot all the students' results and pass, say, everyone who achieved a mark higher than 0.5 standard deviations below the mean, you could use this method every year to find the top 69 per cent of students.

CHAPTER 15

How long is a piece of string?

Not everything can be counted.

Counting is useful for groups of objects, such as cows, cakes, saucepans or sheep. But not everything is in discrete chunks. Instead, we measure continuities, such as time, fluids, and things we can't be bothered to count, such as grains of sand or rice.

Rulers and rulers

The earliest units of measurement we know about were based on the human body: the length of a pace, the distance from fingertips to elbow (a cubit), or the length of the top joint of the thumb. These are good enough as long as you are not working on anything very precise or ambitious, and don't need to combine components made or measured by different people. But imagine trying to build a great pyramid if each side was paced out by a different person. Even the same person might take different-length paces on each side. In this situation *standardization* rapidly becomes useful.

Having settled on one person's arm – say, the pharaoh's or the chief architect's – to measure a cubit, it would be inconvenient for that person to have to hang around all the time just for measuring purposes, and anyway they can't be in more than one place at a time. A stand-in – such as a ruler – would be useful here. The royal cubit was often represented by a wooden rod, marked into divisions, just like a modern ruler. This type of standardization, beginning 5,000 years ago, worked well: the Great Pyramid at Giza is built on a base 440 cubits square, accurate to 0.05 per cent, or 115mm (4.5in) in 230.5m (756ft).

SI units

The metric and decimal system that forms the basis of the SI (Système International) units began in France in 1799. Now most of the world uses SI units.

There are seven base SI units set out by the 11th General Conference on Weights and Measures in 1960:

- **ampere** (A) – unit of measurement of electric current
- **kilogram** (kg) – unit of measurement of mass
- **metre** (m) – unit of measurement of length
- **second** (s) – unit of measurement of time
- **kelvin** (K) – unit of measurement of thermodynamic temperature (a kelvin is equal to one degree Celsius, but the starting point is at absolute zero, the equivalent of minus 273.15° Celsius)
- **candela** (cd) – the unit of measurement for luminosity
- **mole** (mol) – the amount of any substance that contains the same number of elementary particles (such as atoms, ions or molecules) as 12g of carbon-12. This equates to Avogardo's constant: $6.02214129(27) \times 10^{23}$ atoms/molecules.

MEASURES FOR ALL

People around the world developed different measuring systems according to what they needed to measure. As a result there are some quite odd units of measurement, including:

- a horse length (2.4m/7.9ft) in horse racing
- a cow's grass (unit of area) – enough land, if laid to grass, to sustain one cow
- a morgen (unit of area) – as much land as can be tilled by one man and an ox in a morning, defined by the South African Law Society in 2007 as equal to 0.856532 hectare/2.1 acres (which seems a bit too precise)
- the mass of Jupiter – used to report the mass of exoplanets, equal to 1.9×10^{27}kg.

There are many more SI units which are defined in terms of these base units. Some familiar measures, such as hour, litre and tonne, are not SI units.

Twenty officially sanctioned prefixes are used with SI units:

Factor	Name	Symbol
10^{24}	yotta	Y
10^{21}	zetta	Z
10^{18}	exa	E
10^{15}	peta	P
10^{12}	tera	T
10^{9}	giga	G
10^{6}	mega	M
10^{3}	kilo	k
10^{2}	hecto	h
10^{1}	deka	da

Factor	Name	Symbol
10^{-1}	deci	d
10^{-2}	centi	c
10^{-3}	milli	m
10^{-6}	micro	μ
10^{-9}	nano	n
10^{-12}	pico	p
10^{-15}	femto	f
10^{-18}	atto	a
10^{-21}	zepto	z
10^{-24}	yocto	y

In practice, we don't measure time in megaseconds but in months and years.

ROGUE MEASURE

In 2001, US student Austin Sendek proposed the prefix 'hella' to denote one octillion (10^{27}) of an SI unit. The Consultative Committee for Units considered and rejected the proposal, but it has since been adopted by some websites, including Google Calculator.

How standard is a standard?

Tools for measuring must be calibrated against a defined standard, which therefore must be absolutely invariable. That sounds straightforward, but it is not. A wooden rule can shrink and distort as the wood dries out; even an iron rod will expand in the heat and contract in the cold.

Today, only the kilogram is still based on a human-made, physical standard. The other SI units are based on immutable features of the universe. For example, the duration of a second is '9 192 631 770 periods of the radiation corresponding to the transition between the two hyperfine levels of the ground state of the caesium 133 atom'.

So, how long?

The units of measurement we use most as individuals are probably those for length or distance. Most of us might deal with measurements from a few millimetres to hundreds or even thousands of kilometres, so we use millimetres, centimetres, metres and kilometres. But this is only a tiny portion of the full range.

Defining the metre

The metre was first defined as $1/_{10,000,000}$ part of half a meridian (that is, half the distance around the Earth from the North Pole to the South Pole) as measured in 1795. This was accurate to within about half a millimetre. It was represented by a standard in Paris in the form of a platinum bar, itself accurate to within about a hundredth of a millimetre. It switched to a non-physical standard in 1960, and is now defined as the distance travelled by light in a vacuum in 1/299,792,458 of a second – which suggests it might have been better to redefine the length of the metre to the path travelled by light in $1/_{300,000,000}$ of a second, but we'd already gone rather a long way with the old metre by then.

If a piece of string is centimetres, metres or even kilometres long, that's fine, but if it were to extend from here to Neptune, we'd be more likely to measure it in AUs – *Astronomical Units* (not an SI unit). An AU is the average distance from the centre of the Earth to the centre of the Sun, or 149,597,870,700m (about 93 million miles).

THE METRE'S NOT GOOD ENOUGH

When Anders Ångström developed the unit in 1868, the standard for the metre was a platinum bar kept in Paris. Dealing with a unit so small that the distance between atoms can be measured in it, a metal bar is not the best standard – what if a few extra atoms were stuck on the end? Early on, Ångström had made an error of about one part in 6,000 and so had his metal bar checked against that in Paris. The comparison was not very accurate, and his corrected calculations were worse than the originals. In 1907, the ångström was redefined with the wavelength of the red line of cadmium in air equal to 6438.46963 ångströms.

Outside the solar system, units of measurement become ever larger. Astronomers use units that would be completely useless on Earth.

A *light year* (ly) is the distance that light travels in a year: 9,460,000,000,000km (5.88 trillion miles). Measuring in light years is of limited use within the solar system. Better to measure in *light minutes* (the distance light travels in a minute) and *light hours*. The Earth is 499 light seconds from the sun, which means it takes light 8 minutes and 19 seconds to reach the Earth. If the sun exploded right now, you would have just over eight minutes of blissful ignorance before you knew about it. Neptune is 30AU, or 4.1 light hours, from the sun.

Astronomers don't really like light years – they are not a very scientific-sounding measure. They prefer to use parsecs. The name comes from 'parallax of one arc second'. A parsec is 3.26 light years or 206,265 AU.

Although we don't talk about kiloAU or kilolightyears, we do talk about kilo- and megaparsecs. A *megaparsec* is a million parsecs, or around 200 billion times the distance from the Earth to the Sun, and a *gigaparsec* is a billion parsecs. The diameter of the observable universe is thought to be about 28

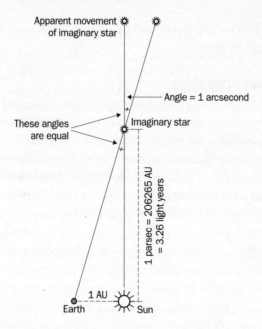

Apparent movement of imaginary star

Angle = 1 arcsecond

These angles are equal

Imaginary star

1 parsec = 206265 AU = 3.26 light years

1 AU

Earth   Sun

gigaparsecs. It seems unlikely we will ever need a unit of measurement larger than the gigaparsec. A piece of string is not going to be too long to measure in gigaparsecs.

Or how short?

If the piece of string is super-short, we might measure it in ångströms (Å); 1Å is 10^{-10}m, or one ten-billionth of a metre. The distance between the centres of two carbon atoms in a diamond is about 1.5 Å.

Most of an atom is empty space. Although a carbon atom might be 1.5 Å across, the protons and neutrons in the nucleus are around 1.6×10^{-15}m, or 1.6 femtometres, across. The rest of the space is patrolled rather randomly by electrons. An electron is thought to be between 2×10^{-15}m and 10^{-16}m across, though it's a bit of a fudge as an electron is also said to have no definite spatial extent – that is, it doesn't take up any space. If you had a ruler marked in femtometres, you could go around measuring atomic nuclei and electrons.

Short and super-short

Great – but what if you had a ruler that was just a thousandth of a femtometre long, a one-attometre ruler, marked with zeptometres (a thousandth of an attometre)? What could you possibly measure with it? You could measure the larger quarks (a type of subatomic particle), but top quarks are smaller than a zeptometre (10^{-21}m), so you'd need a new ruler, perhaps a zeptometre rule divided into yoctometres. (If you think of the femtometre rule as being a kilometre long, the yoctometre would be a millionth of a millimetre – and the femtometre is itself smaller than the nucleus of an atom, remember.)

Now we can measure a *neutrino* (another type of subatomic particle), which is only one yoctometre across (10^{-24}m). Again, the particle doesn't really occupy space in the normal way, but that's the radius of the space over which its force acts. (Think of the way we measure the width of a hurricane: there is no physical object that is a hurricane, but we are content to think of the area in which it is operating as defining its size.) A neutrino is a billionth the size of an electron, so if a neutrino were the size of an apple, an electron would be about the size of Saturn, or ten times the size of the Earth.

At the end of the scale

There is no known particle smaller than a yoctometre – and yet there is a smaller unit of measure. The Planck length is thought to be the smallest unit of length that there can ever be. Although theoretically we could carry on making up smaller and smaller units, they would have no practical use. Below the size of the Planck length – which is 10^{-35}m – the laws of physics don't apply, so measurement itself becomes impossible. The only things that might be measured in Planck lengths are quantum foam and strings in the realms of theoretical physics (if they exist). If an apple were one Planck length in diameter, an electron would be more than 10 million light years across, and a carbon atom would be bigger than the observable universe.

Strings and things

A theory in modern physics suggests that everything – all subatomic particles and therefore everything built up from them – is made of tiny vibrating strings of energy. These strings are tiny – really, really tiny. They are measured in Planck lengths. If a single hydrogen atom was the size of the observable universe, a string would be the size of a tree. So 'How long is a piece of string?' would be better rephrased as 'How short is a piece of string?'

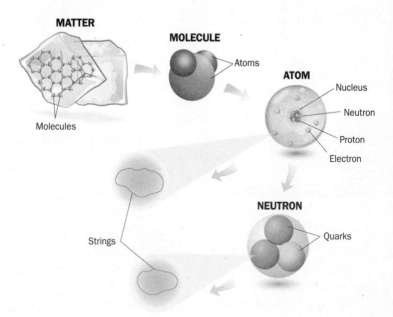

MATTER

Molecules

MOLECULE

Atoms

ATOM

Nucleus

Neutron

Proton

Electron

Strings

NEUTRON

Quarks

CHAPTER 16

How right is your answer?

You wouldn't measure a whale in millimetres and you wouldn't measure an atom in kilometres.

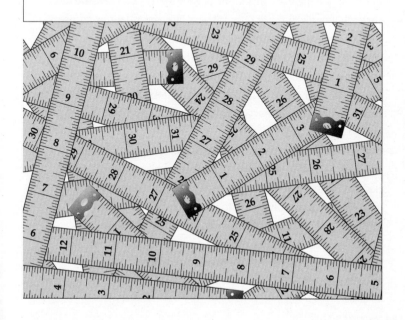

We have lots of different units of measurement (see Chapter 15) so that we can choose one that's appropriate to the thing we're measuring.

Pick a unit . . .

If a unit is too big for the item you are measuring, you will end up either with silly decimals or inaccuracy. Suppose a dog is 69cm (27in) high. This is a good unit. We would not want to say the dog was 0.0069km high.

The volume of the Pacific Ocean is around 660 million cubic km (158 million cubic miles) – but we don't buy milk in cubic kilometres (or miles). As a guide, if there are lots of zeroes before or after the interesting (significant) digits, you should probably pick a different unit.

Counting and calculating

Counting is pretty straightforward – we can easily count how many people there are in a room, or how many cars there are in a car park. But it's difficult to count very large numbers, or numbers that are not stable, or numbers that are not firmly bounded. You couldn't count the grains of sand on a beach for three reasons: there are too many of them, the number changes with the tide and traffic over the beach, and there is no definitive edge to the beach. Where do you start and stop counting? And how deep do you go and still call it 'beach'?

To give a number in these circumstances, we can calculate or estimate. If a multi-storey car park were known to be full, and it had ten floors all of the same design, we could count the cars on one floor and multiply by ten to find out how many cars are in it. The answer is likely to be very accurate. If there were 80 spaces on each floor, the full car park could hold 800 cars – or perhaps on any particular day, 798 or 799 if one or two cars were badly parked.

Calculating and estimating

What about sweets in a jar? Guessing this is a common challenge at a fete or fair.

It's easiest if the sweets are all the same size and shape (preferably spheres or cubes), and the jar is the same width all the way up. The number of sweets in total is going to be close to the number of sweets on one layer times the number of layers of sweets stacked top to bottom. Don't worry about converting to a regular unit; the sweet is the best unit to use here.

For a round jar, count (or guess, if you don't want to look like a nerd or a cheat) the number of sweets in a column top to bottom and the number of sweets in a row going around the jar (or halfway round and double it). Then use this formula, where h is the height in sweets and c is the circumference of the jar in sweets:

$(c/2)^2 \times \pi \times h =$ volume in sweets

It's harder to get a good estimate if the sweets are a jumble of sizes and shapes (polydiverse particles, in science-speak), or if the jar is quite small or an unusual shape. There are ways of calculating it, worked out by scientists who have developed a method for calculating packing density from the point of view of one of the sweets (particles, in their experiments), but that's going a bit far just for a game at a fete. You can make a fair estimate by counting several rows and columns of sweets and taking the mean to use in the formula.

(This game is probably on the way out anyway – you can now get a phone app to calculate the number of sweets in a jar.)

Sampling

At least the sweets aren't climbing in and out of the jar or moving around inside it. Nor are they hiding from you. What if you wanted to calculate how many rooks live in a stand of trees? They come and go, they hide in their nests, and there are quite a lot of them. The best method might be to observe a sample of trees over time and extrapolate from it, multiplying the estimated number of rooks observed by the estimated number of trees.

Sampling is the method used by polls to predict how people will vote, or to estimate figures such as alcohol consumption or distance commuted. To give a result which can be considered reliable and statistically useful, an estimate based on sampling must use a representative sample of a suitable size. If you wanted to estimate the number of vegetarians in Canada, you would not get a reliable answer if your sample was 15 elderly people in a care home, nor if it was 100 young women on a university campus.

Find the right people

To be representative, a sample must be large enough and diverse enough to reflect the way the population is made up. So, to represent the population of Canada, a survey must include men and women of all ages, ethnicities and socioeconomic groups in about the same proportions as they are found in the whole of Canada. This is called *demographics*.

Working out the right sample size is quite a technical process, which you would need to know about if you were actually going to carry out a survey yourself.

If you are just reading the results of a survey in the media, look out for the sample size and demographic to give you a rough idea of how reliable the results might be. On the whole, the larger the proportion of people sampled, the more the results can be trusted – but only if the researchers have taken care to find a representative sample.

A representative sample

The table on page 142 shows roughly how confident you can be in the results for different sample sizes and populations. For example, if you have a population of more than a million people (as you would have in Canada), to gain a result with a margin of error of just 1 per cent (that is, within ±1 per cent accuracy in the answer) you would need to ask 9,513 people. You could then have 99 per cent confidence in the result.

Again, using a representative sample is crucial. If you wanted to find out about the dietary habits of the population of Canada, a sample of Hindus (mostly vegetarian) or lumberjacks (mostly meat-eaters) would not give a reliable answer.

Population	Margin of error			Confidence level		
	10%	5%	1%	90%	95%	99%
100	50	80	99	74	80	88
500	81	218	476	176	218	286
1,000	88	278	906	215	278	400
10,000	96	370	4,900	264	370	623
100,000	96	383	8,763	270	383	660
1,000,000+	97	384	9,513	271	384	664

Significant figures

One mark of clumsy processing and reporting of statistics is suggesting a false level of accuracy by giving more *significant figures* than is reasonable. Significant figures are the digits that have meaning, showing genuine detail in a number – they don't include zeroes that are just placeholders. The number 103.75 has five significant figures, the most significant of which is 1, as it shows the number is 100 and something. The number 121,000 probably has only three significant figures– unless the number indicated is *exactly* 121,000.

We limit significant figures when rounding up or down. When the result of a calculation is not likely to be very accurate, it's reasonable to round it to the figures that are likely to be correct. For example, if you counted the grains of sand in a teaspoon and then calculated that there were 445,341,909 grains in a container of sand, this is precise but not likely to be very accurate. It would be reasonable to round it up to 450,000,000 or down to 400,000,000 as you can't be accurate to within 50,000 in a calculation like this. The population of the world, which is impossible to measure accurately and is constantly changing, is often given as 7 billion. A current estimate (in 2015) is 7,324,782,000. A more precise figure would not be more accurate and would suggest we know more than we do.

Sometimes, a calculation yields a result with more significant figures than is appropriate to show. Suppose you wanted to know the area of a circular rug with diameter 120cm (47in). The expression giving the area of a circle is πr^2, so for a circle of radius 60cm (23.5in), the area is 11,309.7336cm². For most purposes, 11,300cm² (4,450in²) would be precise enough. In any case, it's not appropriate to include the figures after the decimal point as the radius of the rug wasn't measured with that level of accuracy in the first place. To include them suggests a greater degree of precision than has actually been achieved.

PRECISELY PI?

Pi (symbol π) is an irrational number, which means that the digits after the decimal point go on and on in an unending sequence.

Although computers have now calculated π to billions of places, mathematicians believe there is usually no point in using more than 39 decimal places as that's sufficient to calculate the volume of the known universe to the size of one atom.

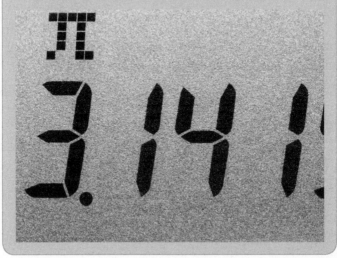

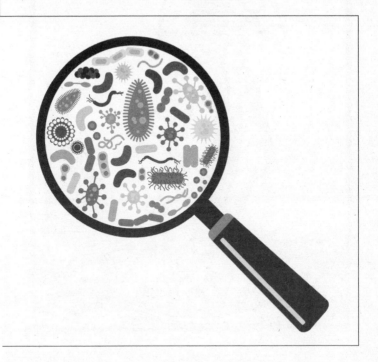

CHAPTER 17

Are we all going to die?

Pandemics are scary things.

A pandemic is an epidemic that spreads over continents or even the whole world.

A plague on all your houses

The most famous pandemic is the Black Death, which in 1346–50 killed up to 50 million people in Asia, Europe and Africa. Most medical historians think it was a particularly vicious form of *Yersinia pestis*, the bacterium that causes bubonic plague. The next major pandemic was caused by a new strain of flu in 1918–19. It spread around the world, killing 50–100 million people. The numbers are similar to those killed by the Black Death, but the world population was much larger in 1918 (nearly two billion) than in 1346 (around 400 million). Could it happen again?

Should we be scared?

It's quite reassuring that global pandemics on this scale have happened only twice. But the world has changed over the last 100 years. With modern patterns and speeds of international travel, the Black Death would take only weeks or months to spread around the globe, rather than the years it took in the Middle Ages, when no one could move faster than a horse could run (or, more usually, trudge). The mathematics are very different now.

A pathogen's guide to success

Epidemic and pandemic diseases such as flu and bubonic plague are caused by pathogens – often bacteria or viruses. To cause a pandemic, a pathogen needs to:

- be easily transmitted between people
- be transmissible before people are so ill they can't go out and make contact with other potential victims
- let people live long enough to pass it on.

Ideally, a pathogen needs to know some mathematics so that it gets it all right.

Again and again and again – reproduction rate

The critical number in determining whether a pandemic can occur is the basic reproduction number of the disease, known as R_0. It is a calculation of how many other people a single, typical case will infect during the time they are infectious – that is until they either die or fight it off and are no longer infectious. The higher the R_0, the much better the chance the pathogen has of causing a pandemic. In a simple model, if the $R_0 < 1$, there will be no epidemic and if $R_0 > 1$ there will be an epidemic, although in practice it is a bit more complicated. R_0 can be calculated by gathering data about individual cases and tracing their contacts and infection rates, or by gathering data about infection rates in a whole population. The two methods often give quite different results, which makes *epidemiology* (the study of epidemics) a challenge.

To calculate R_0:

$$R_0 = \tau \times \bar{c} \times d$$

τ is the transmissibility – that is, the probability of infection when an infected person is in contact with a susceptible person. If an infected person has contact with four people and one of them becomes infected, the transmissibility is 1 in 4, or ¼.

$\bar{c}$ is the average rate of contact between susceptible and infected individuals, calculated as contacts divided by time. If there are 70 contacts between an infected person and a susceptible person over a week, the contact rate per day is $^{70}/_7 = 10$.

d is the duration of infectiousness – how long someone remains infectious (in the same time unit as $\bar{c}$ has been calculated).

If a disease makes people infectious for four days, the transmissibility is ¼ and the rate of contact is 10, then

$$R_0 = \text{¼} \times 10 \times 4 = 10$$

This pathogen has a good chance of infecting a lot of people!

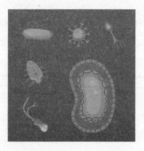

Another important factor is how many people are susceptible. People are not susceptible if they are immune, either because they have had a particular disease before or have been vaccinated against it. With a new strain or disease, everyone is likely to be susceptible, making it much easier for the disease to spread.

As a general rule, the higher the R_0 value, the more difficult it is to control the spread of a disease. As there are many ways of calculating R_0, some in the field and some theoretical, the numbers are not very reliable or necessarily directly comparable. But they are the best we have – the pathogens have the upper hand in this one.

We are not numbers

R_0 can only ever be an approximation. It's usually based on the assumption that the population and the number of contacts within it are *homogenous* (evenly mixed). But that's rarely true in reality because some people will be more vulnerable than others. For example, some people will mix with a large but non-homogenous group, such as teachers who are in contact with groups of children, or elderly people who live in a care home. Those who live alone or in a remote community will have only limited contact with others.

All change

The R_0 value of a disease changes during the course of an epidemic or pandemic. Two of the values in the calculation, transmissibility and contact rate, depend on the number of susceptible people. This decreases as the epidemic continues, since the number of susceptible people will go down; people are no longer susceptible once they have caught the disease and recovered (or if they have died).

To start with, all the contacts of an infected person are susceptible (in an unvaccinated population):

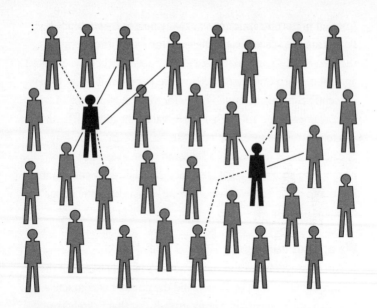

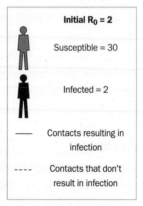

Initial $R_0 = 2$

Susceptible = 30

Infected = 2

—— Contacts resulting in infection

---- Contacts that don't result in infection

LOOK OUT – DISEASES ABOUT

The approximate R_0 values of some common epidemic diseases are:

Disease	R_0
Measles	12–18
Pertussis (whooping cough)	12–17
Diphtheria	6–7
Polio	5–7
SARS	2–5
Influenza (1918 pandemic)	2–3
Ebola (2014 outbreak)	1.5-2.5

Later on in the epidemic, many of the contacts have already had the disease, so are no longer susceptible:

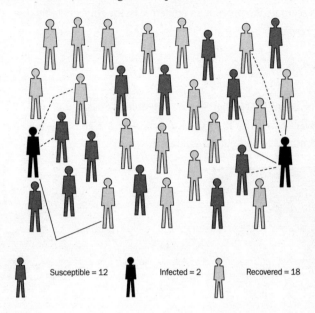

Susceptible = 12   Infected = 2   Recovered = 18

The R_0 late in the epidemic will be lower. Eventually it will drop below one and the epidemic will end.

Get a jab!

Vaccines work by reducing the number of susceptible people in a population. If most people are vaccinated, the chances of an infectious person coming into contact with a susceptible person are low, so an epidemic can't start. This is called *herd immunity* and it helps protect those who can't have vaccinations (because they have cancer or HIV, for instance). It works by reducing the chance that they will come into contact with an infected person because most people they meet will be immune. If they do encounter the disease, they will have no personal protection against it, so the greater the herd immunity the safer they are.

If a vaccine were 100 per cent effective in preventing disease, the proportion of people in a population who would need to be vaccinated to prevent an epidemic is, roughly, given by the expression

$1 - 1/R_0$

This means that if there were a threatened epidemic of a deadly flu with an R_0 of 3, $1 - 1/3 = 2/3$ of people would have to be vaccinated to prevent the epidemic.

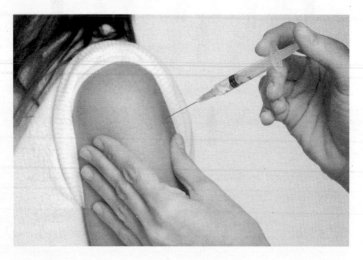

Measles has an R_0 of 12–18. Let's split the difference and call it 15 for simplicity. That means that $1 - 1/15 = 14/15$ or 93 per cent of people must be vaccinated to stop measles spreading in a population. Around 20 per cent of Americans wrongly believe vaccination can cause autism, and some refuse to have their children vaccinated for this reason. In the USA, the national uptake in 2015 was 91.1 per cent, but in some areas it was as low as 81 per cent for pre-school nursery children, making those areas more vulnerable to a measles epidemic.

CHAPTER 18

Where are the aliens?

Surely we are not the only intelligent life forms in the universe?

Just looking at our corner of the universe, there are thought to be 300–400 billion stars in the Milky Way. Our galaxy is not even very big: giant elliptical galaxies have around 100 trillion stars each. With (probably) more than 170 billion galaxies in the observable universe, there could be 10^{22} to 10^{24} stars. Even 10^{22} stars is 10,000 stars for every grain of sand on all the beaches in the world, and 10^{24} is 1,000,000 stars for every grain of sand. It would be incredibly arrogant to assume that we are so special that there are no other technologically advanced civilizations in a universe of 10^{22} stars.

Observable universe

The observable universe is a sphere centred on Earth and around 92 billion light years across. There might be a lot of universe beyond that, but we can't know about it as any light leaving it won't have reached us yet, even after 13.8 billion years. There's likely to be more universe we don't know about. The probability that we, by coincidence, are right in the middle of a spherical universe is tiny.

So it seems highly likely that intelligent beings exist elsewhere in the universe, but they might be too far away to contact us, even if they wanted to. But what about the probability of intelligent life among the 300–400 billion stars that make up our own galaxy? This is something we may be able to answer – one day.

The Fermi paradox

The Italian physicist Enrico Fermi (1901–54) remarked in 1950 that if intelligence is common in the universe, why have we not had any contact with or seen any evidence of aliens? It's a question that has puzzled astronomers ever since, and prompted many theories about barriers to technological development or to species evolution and survival, as well as reinforcing the old question of whether we are actually rather special.

Enrico Fermi became famous for creating the nuclear reactor. He made his comment about aliens during a casual lunchtime chat. Since then the search for alien life in our universe has – well – sky-rocketed.

SIZE AND THE SPEED OF LIGHT

The radius of the observable universe is greater than 13.8 billion light years, even though the universe is reckoned to be 13.8 billion years old. This is because the expansion of space has pushed the furthest bits further away during all that time. Light that set out on its journey to us 13.8 billion years ago has come from objects that are now around 46 billion light years away.

The Drake equation

The Drake equation attempts to set paramaters for the likelihood of intelligent life outside Earth, but within our own galaxy. We don't yet have the data to fill in all the variables, but it shows how the probability calculation could be made if we had the right data.

> 'Intelligent life in the universe? Guaranteed. Intelligent life in our galaxy? So overwhelmingly likely that I'd give you almost any odds you'd like.'
>
> Paul Horowitz, leader of SETI (Search for Extra-Terrestrial Intelligence, 1996)

There are a few slightly different versions. The more intuitive version looks like this:

$$N = N* \times f_p \times n_e \times f_l \times f_i \times f_c \times f_L$$

where:

N = the number of civilizations whose electromagnetic emissions are detectable in our galaxy (so those which are on our current light cone, see diagram below)

and

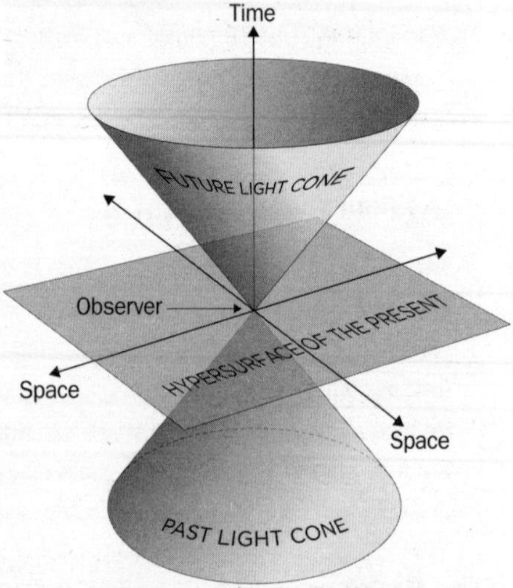

N∗ = the number of stars in the Milky Way

f_p = the fraction of those stars which have planets

n_e = the average number of planets per solar system which can potentially support life

f_l = the fraction of planets which could support life that actually develop life at some point

f_i = the fraction of planets with life which go on to develop intelligent life (civilizations)

f_c = the fraction of civilizations that develop a technology which releases detectable signs of their existence into space

f_L = the fraction of the planet's life during which the communicating civilizations live.

Although the odds look stacked against intelligent life, remember that we are starting with 300 or 400 billion stars in our galaxy and that it seems increasingly likely that *having* planets is the norm, rather than the exception. Let's experiment with some numbers, all hypothetical.

Let's say 15 per cent of the stars in the galaxy are sun-like and could have planets (f_p). This is around the middle of the current range of estimates, 5–22 per cent.

400 billion x 0.15

In our solar system of planets, in addition to Earth, Mars is the only planet thought to have once sustained life or to have had the potential to do so, so we will make n_e = 2:

400 billion x 0.15 x 2

PLANETMANIA

Until recently, we had no idea whether any other stars in the Milky Way had planets. But now the search for exoplanets (planets outside our own solar system) is well underway and turning up lots. By April 2015, more than 1,900 exoplanets were known, in over 1,200 different planetary systems.

Many scientists believe that life started on Earth after only a billion years or so. Does that mean life is very likely to start if the conditions are right? But we haven't found life on other planets and moons in the solar system which look as though they could support life, so this suggests that maybe it does not appear so readily. We don't know.

Estimates of how likely the emergence of life may be range from 100 per cent (if life can emerge, it will) to close to 0 (it's very rare for life to emerge). Let's pick a figure in the middle and say 10 per cent (f_l):

400 billion x 0.15 x 2 x 0.1

How many of those planets will have life forms which evolve intelligence (f_i)? This is very hard to guess. Some scientists think intelligence is so advantageous that it will eventually emerge, so close to 100 per cent – others think it very rare. Let's pick 1 per cent:

400 billion x 0.15 x 2 x 0.1 x 0.01

Now it becomes highly speculative. We have no idea how likely it is that an intelligent species will build a technological civilization and produce detectable electromagnetic signals (f_c). It could be 1 in 10 or it could be 1 in a million. Let's go for 1 in 10,000.

400 billion x 0.15 x 2 x 0.1 x 0.01 x 0.0001 = 12,000

So we have 12,000 civilizations in the Milky Way capable of producing signals we can detect. That looks promising, but most importantly they need to overlap with us in time – or rather, the arrival of their transmissions needs to overlap with us in time.

If a civilization remains capable of electromagnetic activity for 10,000 years (the duration of human civilization so far), and its planet lasts 10 billion years, then f_L is: $10^3 \div 10^9 = 1/10^{-6}$

12,000 x 10^{-6} = 0.012

So there is a 98.8 per cent chance that no one is out there listening or beaming out signals in our galaxy just now.

Of course, all our figures are highly speculative and might be completely wrong. If half of all stars have planets that could support life, if life were sure to emerge and eventually become intelligent, if 10 per cent of intelligent life develops electromagnetic communications and if the most successful species, like sharks, survive 350 million years, the figures are very different: we now have 14 billion communicating life forms! This is more than a trillion times as many as the more conservative figures suggest.

There are various interactive Drake equation calculators online if you want to try different ways of populating the universe.

CHAPTER 19

What's special about prime numbers?

Prime numbers are more useful than you might think, considering that they don't really want to participate in mathematics at all.

1	2	3	4	5	6	7	8	9	10
11	12	13	14	15	16	17	18	19	20
21	22	23	24	25	26	27	28	29	30
31	32	33	34	35	36	37	38	39	40
41	42	43	44	45	46	47	48	49	50

Prime numbers are those that have no factors other than themselves and 1. This means that a prime number is not the product of any multiplication sum (involving only positive whole numbers) except:

[prime number] x 1 = [prime number]

Primes and composites

Composite numbers are numbers which have factors other than themselves and 1. So all positive whole numbers other than 0 and 1 are either prime numbers or composite numbers. Every composite number can be expressed as the product of prime factors, meaning that it can be broken down into a multiplication sum involving only prime numbers. This hints at the importance of prime numbers: they are the building blocks from which we can make all numbers.

SPECIAL CASES

Zero and 1 are not considered to be prime numbers. For a time during the 19th century, many mathematicians did consider 1 to be prime, but it is no longer allowed in the club.

Two is the only even prime.

Prime number theorem

The prime number theorem, proven in the 19th century, states that the probability that a given, randomly chosen number, n, is prime is inversely proportional to its number of digits, or to the logarithm of n. This means that the larger the number, the less likely it is to be prime.

The average gap between consecutive prime numbers up to n is roughly the logarithm of n, or $\ln(n)$.

Finding primes

One test for being a prime number – called 'primality' – is trial division. If n is the number being investigated, try dividing it by all numbers greater than 1 and smaller than $\frac{1}{2}n$.

This is laborious for large numbers, and different methods are used, generally by computer. The largest prime so far discovered (as of April 2015) has 17,425,170 digits, and is $2^{57,885,161} - 1$. It's not worth staying up late trying to find more unless you are very dedicated, but the Electronic Frontier Foundation is offering a prize for the first prime with at least 100 million digits, and also for the first prime with at least half a billion digits.

Some of the greatest mathematical brains, and now also the most sophisticated computer programs, have searched for patterns in the primes, but no predictable pattern has so far been found.

The sieve of Eratosthenes

The Ancient Greek mathematician Euclid of Alexandria, living in the 2nd or 3rd century BC, was the first person we know of to have recognized prime numbers. Another Greek mathematician, Eratosthenes, of the 2nd century BC, introduced his so-called 'sieve' for identifying prime numbers. It is only feasible for relatively small numbers, but it is easy to use.

Draw up a grid with 10 columns and as many rows as you need to accommodate the numbers you want to check: if you want to check up to n, you need a grid showing 1 to n. Starting with 4, go through the grid and cross out all multiples of 2. Then cross out all multiples of 3, then of 5, then of 7, and so on, working your way through the primes. When you have got as far as multiples of $\frac{1}{2} n - 1$, you can stop, as larger numbers can't be factors of n or less. The numbers that are not crossed out are primes.

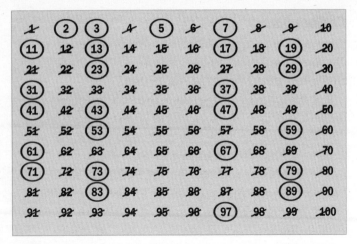

To discover whether a number is a prime, try dividing it by 2. If this gives you a whole number, it can't be a prime. The only even prime number is 2, which if divided by 2 gives you 1 (not a prime number).

Sadly neglected

Between the Ancient Greeks and the 17th century there was little interest in primes. Even in the 17th century, primes had no real uses outside pure mathematics. They came into their own in the computer age, with the need to develop encryption algorithms.

Putting primes to work

Primes had a pretty lazy time of it until the need for data encryption came along. Now we send gazillions of secure transactions and other secret data across the internet daily, primes provide the equivalent of the Securicor vans that data travels in.

Start by multiplying two very large prime numbers together to give a composite number:

$$P_1 \times P_2 = C$$

The composite number is used to generate a code called a public key, which the bank (or whatever) sends to the person who wants to encrypt their data. If you were buying something online, your credit card details would be encrypted using this public key, the encryption happening at your end of the connection. The encrypted data would be gobbledegook if intercepted in transit. When your card details arrive at the other end, the private key – made from P_1 and P_2 – is used to unencrypt the data.

This works because it's very difficult to find the prime composites of large numbers. Any hacker who intercepted the data would need 1,000 years of computer time to crack the code and find the original primes. It's because it's so hard to crack modern encryption that governments would really rather like tech companies to build 'backdoors' into their systems so that it's easier for them to see what people are doing.

ULAM SPIRAL

In 1963, Stanislaw Ulam was idly doodling during a boring scientific presentation when he made an astonishing discovery. He drew a spiral of numbers, with 1 at the centre.

```
37—36—35—34—33—32—31
 |                   |
38  17—16—15—14—13  30
 |   |            |   |
39  18   5— 4— 3  12  29
 |   |   |     |  |   |
40  19   6   1— 2  11  28
 |   |   |        |   |
41  20   7— 8— 9—10  27
 |   |               |
42  21—22—23—24—25—26
 |
43—44—45—46—47—48—49...
```

He then isolated all the primes (below).

He noticed a tendency for the primes to fall on diagonals. The larger the spiral, the more obvious the pattern becomes. Some fall on horizontal or vertical lines, too, but not to the same extent.

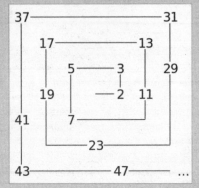

If a computer program is used to plot a white pixel for a composite number and a black pixel for a prime number, in a Ulam spiral, the diagonals appear clearly. Comparing with a plot of the same number of random numbers shows the diagonals really are there.

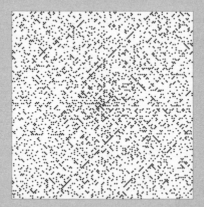

It's still not a predictable pattern, though it's a tantalizing suggestion that there might be some kind of pattern in there somewhere.

'[Primes] grow like weeds among the natural numbers, seeming to obey no other law than that of chance [but also] exhibit stunning regularity [and] that there are laws governing their behaviour, and that they obey these laws with almost military precision.'

Don Zagier, American number theorist (1975)

CHAPTER 20

What are the chances?

We all work with probabilities (in other words, chance or risk) every day, even when we are not aware of it.

When you buy a lottery ticket – or just cross the road – you are dealing with probability.

Against the odds?

Probability is at the heart of gambling. Indeed, it was gambling that prompted the first work in probability, the mathematical face of chance or risk. A casino owner or bookmaker has to understand probability well enough to come out ahead most of the time, otherwise they won't make a profit. But they have to present the odds in a way that makes a bet look attractive to people. There are several ways of doing this.

At the races, bookmakers offers odds on horses expressed as ratios, like this:

Dire Warning	**20:1**
Fancy Pants	**4:1**
Strange Quark	**8:1**
Tangent	**7:1**
Fair Bet	**5:1**

It's easier to see what's going on if we think of these as fractions. The odds 20:1 mean the bookie thinks Dire Warning has a 1 in 20 chance of winning, or $\frac{1}{20}$. Fancy Pants has a much better chance, at $\frac{1}{4}$. If we added up the odds for all the horses, as fractions, the total should, in terms of mathematics, come to 1. It never does, of course, because the bookmakers make their profit out of the difference between the total bets and the total payouts. In this race, the total is:

$\frac{1}{20} + \frac{1}{4} + \frac{1}{8} + \frac{1}{7} + \frac{1}{5}$
$= 0.05 + 0.25 + 0.125 + 0.142857 + 0.2$
$= 0.767857$

The difference between this and 1 is 0.232143, meaning that the bookie makes just over 23 per cent profit (assuming bets are evenly placed).

Clearly, the bookie's odds don't relate to the real odds – the bookie has understated each horse's chance of winning. The real probabilities must add up to 1 as there will certainly be one horse that wins (unless *all* the horses fall or are disqualified).

A lot on the lottery

Other types of gambling offer a low chance of a really big win but a reasonably good chance of a small win. Many national lotteries work like this. The chance of winning the jackpot is very small – often 1 in many millions – but there is a much better chance (perhaps as high as 1 in 25) of winning a very small prize, such as £10. This is cunning, as it reassures people who are aware that the chance of winning the jackpot is tiny by showing them that they might not lose their money, at least. Advertising might mention that there are '50,000 prizes every week', or something similar. As we have seen in Chapter 9, the mention of the large number of prizes appeals immediately; denominator neglect means we don't think of it as a chance of 50,000 in (say) 3.5 million, or just 1 in 70.

Fruit machines operate on the same principle of graded payments, so offer a reasonable chance of winning a little, but a low probability of winning a lot. Someone who wins a small sum is encouraged to try again and can end up losing a lot of money.

Again and again

Sometimes, it's useful to know the probability of more than one event happening. We might want to know:

- **the chance of A or B happening**
- **the chance of both A and B happening**

To work out alternative chances (A or B), we add the probabilities together.

To work out cumulative chances (A and B), we multiply probabilities.

Suppose you apply for two jobs. For the first job, there are five equally well qualified applicants (including you), so your chance of getting the job is 1 in 5, or 0.2. For the second job, there are only four qualified applicants, so your chance of getting the job is 1 in 4, or a probability of 0.25.

The chance you will be offered either (or both) of the jobs is:

0.2 + 0.25 = 0.45 (45 per cent)

The chance you will be offered both jobs is:

0.2 x 0.25 = 0.05 (5 per cent)

You are eight times more likely to be offered one job than both.

The chance of you being offered one job, but not both, is the difference between the chance of being offered one or both and the chance of being offered both:

0.45 - 0.05 = 0.4 (40 per cent)

The most likely outcome, then, is that you will get neither, with the next most likely being that you will be offered only one.

More than one way

It's often easiest to see the principles involved in calculating probabilities by thinking about tossing coins and rolling dice.

Tossing a coin is straightforward: it can land either heads or tails. Assuming it's a fair coin and equally likely to fall either way, the chance of getting heads is ½ (0.5, or 50 per cent) and the chance of getting tails is also ½ (0.5, or 50 per cent). If we toss the coin twice, we could again get heads or tails. The possibilities for two coin tosses are:

Toss 1	heads		tails	
Toss 2	heads	tails	heads	tails

There are now four possible outcomes: heads, then heads; heads, then tails; tails, then heads; tails, then tails. For many purposes, heads, then tails is the same as tails, then heads. The chance of heads twice is ¼; the chance of tails twice is ¼; the chance of heads once and tails once is ½.

The possible outcomes increase as we toss the coin more times, and the chance of getting all the same – heads or tails – reduces (see table top right). The chances of getting heads every time for a number, n, of throws is $\frac{1}{2}^n$. The chances of getting tails every time is also $\frac{1}{2}^n$. The chances of getting all heads or all tails is 2 x $\frac{1}{2}^n$, which is the same as $\frac{1}{2}^{n-1}$ (see right).

Number of throws	Chances of all heads
1	½
2	¼
3	⅛
4	1/16
5	1/32
6	1/64

Number of throws	Chances of all heads OR all tails
1	1
2	½
3	¼
4	⅛
5	1/16
6	1/32

1 in 6

With dice, the problem is more complex as there are six possible outcomes for each throw of a die. The same calculation applies, but now with powers of 6, so 6^n. The chances of throwing a five (or any other number) each time are shown in the table below.

If you throw two dice, the chances of throwing any double are 6^{n-1}.

As soon as you have more than one throw, the probabilities for different total scores become more complex. This is because there is more than one way of scoring some numbers, but not others (see table at the bottom of the page).

Number of throws	Chances of all 5s
1	$1/6$
2	$1/36$
3	$1/216$
4	$1/1,296$
5	$1/7,776$
6	$1/46,656$

The number you are most likely to get if you throw two dice is 7, as there are six ways of making 7. This means the probability of scoring 7 is $6/36$ or $1/6$. If you have a choice in a game of dice, opt for needing to throw 7.

2	1 + 1					
3	1 + 2	2 + 1				
4	2 + 2	1 + 3	3 + 1			
5	1 + 4	2 + 3	3 + 2	4 + 1		
6	1 + 5	2 + 4	3 + 3	4 + 2	5 + 1	
7	1 + 6	2 + 5	3 + 4	4 + 3	5 + 2	6 + 1
8	2 + 6	3 + 5	4 + 4	5 + 3	6 + 2	
9	3 + 6	4 + 5	5 + 4	6 + 3		
10	4 + 6	5 + 5	6 + 4			
11	5 + 6	6 + 5				
12	6 + 6					

HELPING YOU DECIDE

The 19th-century psychiatrist Sigmund Freud used to encourage people who were feeling indecisive to toss a coin to help with difficult life decisions. He wasn't advocating leaving important choices to chance, but using the coin to help identify desires:

'What I want you to do is to note what the coin indicates. Then look into your own reactions. Ask yourself: Am I pleased? Am I disappointed? That will help you to recognize how you really feel about the matter, deep down inside. With that as a basis, you'll then be ready to make up your mind and come to the right decision.'

CHAPTER 21

When's your birthday?

If there are 30 people in a room, there's a good chance (that is, much better than evens) that at least two of them will share a birthday.

The oft-quoted statistic cited on the previous page is hard to believe. It seems totally counter-intuitive.

Frequentist birthdays

There are two ways of working with probability. One of them uses the methods we've looked at in Chapter 20. This is called a frequentist method. The other is a Bayesian method, devised by the English mathematician Thomas Bayes (1702–61), and is more complicated.

There are 365 days in a year (ignoring leap years). So there is a $1/365$ chance of your birthday being on any particular day. If you were comparing yourself with just one other person, then the chance of your birthday matching theirs would be $1/365$

= 0.0027

But don't forget that it's not just your birthday that's interesting. There are 30 people in the room, which gives 30 x 29 possible birthday pairs, or 870. Now you can see why it's so likely there will be a shared birthday.

Inverting the problem

Instead of thinking about the chances of shared birthdays, think about the chances of people not sharing a birthday, of there being no matches in a room of 30 people.

When there are just two people, the chance of their birthdays not being the same is

$1 - 1/365 = {}^{364}/365 = 0.997$

If we add a third person, there are now two used birthdays, so only 363 unused days. The chance that none of their birthdays match is now

$$^{364}/_{365} \times {}^{363}/_{365} = \mathbf{0.992}$$

Add another person, and the chance is

$$^{364}/_{365} \times {}^{363}/_{365} \times {}^{362}/_{365} = \mathbf{0.984}$$

By the time you have got 30 people in the room, the chance of there being no matching birthdays is 0.294 – almost 30 per cent. That means there is a 70 per cent chance that at least two people share a birthday. The point at which the probability reaches 50 per cent is when there are 23 people in the room. By the time there are 57 people in the room, the chance of a match is 99 per cent.

Another inversion

The Bayesian approach to probability is rather different. It can work from one set of probabilities to derive another, related probability.

Bayes' theorem states that

$$P(A|B) = \frac{P(B|A)\,P(A)}{P(B)}$$

where P is probability.

When is it all going to end?

One use of Bayesian probability is to calculate the likely expiry date of humanity. Known as the Doomsday Argument, it was first put forward by the Australian physicist Brandon Carter in 1983. He used a rather low figure of 60 billion humans having been born so far (in 1983) to calculate that there is a 95 per cent likelihood that humanity will not last longer than another 9,120 years (fewer than 9,100 years now, as some of the years have passed since 1983).

BAYESIAN TANKS

During World War II, the Allies tried to assess the production of German tanks by carrying out Bayesian analysis on data from tanks that had been captured or destroyed. They worked out how many moulds had been used in making the 64 wheels on two captured tanks. Then from known data about how many wheels could be made from a mould in a month, they calculated the total number of moulds that would give that proportion of matching wheels in a sample of 64. From that they assessed that the Germans were building 270 Panzer tanks a month in February 1944 – far more than previously estimated. They also used Bayesian methods to calculate the likely number of tanks from the serial numbers on captured tanks – with astonishing accuracy.

Comparing the results of statistical assessment with German records (after the war) revealed that statistics was a far more reliable method of working out military capability than intelligence gathering had been.

CHAPTER 22

Is it a risk worth taking?

'Only those who will risk going too far can possibly find out how far one can go.'

T.S. Eliot

Our perception of risk is very strange, and doesn't always relate sensibly to the mathematics of the risk. It is affected by many psychological factors, such as familiarity or novelty, unknown factors (about the risk), the level of control we feel we have, the rarity of the outcome, the inconvenience involved in avoiding the risk, the immediacy of the danger, and the level of harm that could result.

Live dangerously!

Logically, it would seem that if an activity carried a relatively high risk of death or serious injury, we would shun it – yet many people drive too fast, smoke cigarettes and eat more than is healthy. On the other hand, people in the USA and Europe showed a high level of anxiety about Ebola during the 2014–15 outbreak, even though it was limited to six countries in Africa which most of them would never visit.

Ebola had all the hallmarks of a scary risk:

- **infection carried a better than 50 per cent chance of death**
- **the disease is gruesomely unpleasant**
- **it is unfamiliar to most people**
- **there was a high level of media coverage**
- **people felt out of control as disease strikes randomly (though not so randomly that it will infect someone 5,000km/3,100 miles away)**

There were also plenty of unknowns. Could Ebola escape Africa? Could it become transmissible between people before symptoms developed? Yet the inconvenience involved in avoiding the risk was very low: don't go to Africa and hang around an Ebola clinic or handle dead bodies. Most people could be afraid of Ebola without being in any great danger or put to much inconvenience.

On the other hand, travel by car is a known hazard. It is familiar and we feel in control, even if that feeling is somewhat illusory (we are not in control of other drivers). Of course, there is little media coverage of traffic accidents because they are so common, which indicates a high level of risk. Most people are not afraid to travel by car, and to stop doing so would be hugely inconvenient.

High drama, low risk

The animal most deadly to humans is not, as you might think, a shark, tiger, hippo or anything else large. It's not even a dog. It's a mosquito. Mosquitoes kill over half a million people a year through malaria and other diseases. Yet most people would think walking along a riverbank in Brazil was safer than swimming in the shark-prone sea off the coast of Australia. The chance of drowning is 3,300 times greater than the chance of being killed by a shark, so you should count yourself lucky if you survive in the water long enough even to see a shark.

Putting in the numbers

Numbers showing risk, like most numbers, must have context to be meaningful. Here are two figures that relate to road traffic deaths in the USA:

- **in 1950, 33,186 people died in traffic accidents**
- **in 2013, 32,719 people died in traffic accidents**

It would look as though there has been very little progress in road safety since 1950, which is a depressing thought. But adding more information helps to throw light on these figures. If we look at the population of the USA at those times, we can see a bit more clearly what is happening. In 1950, the population was around 152 million. By 2013 it was 316 million – more than double. If we calculate deaths divided by population, it looks as though there has been a definite improvement.

Date	Deaths	Population	Deaths/ 100,000 people
1950	33,186	152m	21.8
2013	32,719	316m	10.3

But if we now look at the number of miles travelled in motorized vehicles in the two years, the figures take on a completely different appearance.

Date	Deaths	Population	Bn vehicle miles	Deaths/ 100,000 people	Deaths/ 100,000,000 vehicle miles
1950	33,186	152m	458	21.8	7.2
2013	32,719	316m	2,946	10.3	1.1

It was seven times more dangerous to drive in 1950 than in 2013; the risk has fallen by 85 per cent.

ANIMALS MORE LIKELY TO KILL YOU THAN A SHARK

On average, fewer than six people worldwide are killed by sharks each year. You stand a much greater chance of being killed by a:

- snake (70,000 deaths per year)
- dog (60,000 deaths per year)
- bee (50,000 deaths per year)
- hippo (2,900 deaths per year)
- ant (900 deaths per year)
- jellyfish (100–500 deaths per year)

One in a million

Risk analysts call a one-in-a-million chance of dying a 'micromort'. If you are thinking about how to get to town or to work, you could compare the risks of different methods of transport using micromorts to calculate how many kilometres you would have to travel before you were likely to be killed in an accident.

Clearly, taking the train is the safest method and going by motorbike is the most risky.

Transport	Km/micromort
Train	9,656km
Car	370km
Bicycle	32km
Walking	27km
Motorbike	10km

Chronic and acute risk

The risk of falling down the stairs and breaking your neck is an acute risk – it could happen now and kill you immediately. If you walk down the stairs without it happening, the risk has gone (for now) and you have suffered no ill effects, except perhaps a little anxiety.

The risk of developing lung cancer if you smoke is a chronic risk. It builds up over time and although any one cigarette you might smoke this afternoon is not going to kill you, it might – with all the others – contribute to an early death. This risk is cumulative; every cigarette you smoke increases your risk of lung cancer and some other conditions.

Microlife and microdeath

The opposite of a micromort is a microlife – a millionth of a life. For a young adult, this is on average about half an hour. Chronic risks are often better expressed in terms of the cost in microlives. Smoking one cigarette costs about one microlife. Of course, it is not a direct and indisputable cost – it's a risk. If we take the average lifespan of people who smoke a certain number of cigarettes and compare it with the average lifespan of nonsmokers we can work out the average cost in microlives of a cigarette. But some people smoke 20 cigarettes a day and live to be 90; nothing is certain.

The crucial difference between using micromorts to calculate the acute risk of an activity and using microlives to calculate chronic risk is that the cost in microlives is cumulative whereas the risk in micromorts is reset to zero each time you survive.

HOW MUCH IS A LIFE WORTH?

Governments make decisions about spending on safety based on calculating lives that can be saved. They use a figure called Value of a Statistical Life (VSL) or Value for Preventing a Fatality (VPF) to work out the economic value of different life-saving measures. In the UK, a life is worth £1.6 million ($2.4 million) in terms of road improvements, so a microlife is worth £1.60 ($2.40). In the USA, lives are worth more, with the US Department of Transport valuing a VSL at $6.2 million (£4 million) and a microlife at $6.20 (£4).

Everything's a risk

Another way of considering risk is to compare it with a baseline risk that you run just by being alive. The chances of dying in a hang-gliding accident are about 1 in 116,000 for each flight you take. A 30-year-old American male has a 1 in 240,000 chance of dying on any particular day, so he increases his risk three-fold by going hang-gliding (since he adds the new risk to his existing risk – it doesn't replace it).

Another way of representing risk is to show how long you would have to do an activity, continuously, to come to grief, or calculate the risk of each instance of an activity. If the risk of dying on each hang-gliding trip is 1 in 116,000, that suggests that if you set out to go hang-gliding 116,000 times, you would be pretty likely to die at some point (though it might be on your third flight, or your 169th, not your 116,000th). While on average this is true, it's not necessarily true for any particular person. There could be other factors at play. Early hang-gliding flights might be more dangerous because the pilot is inexperienced. Later flights might be more dangerous because the pilot has become blasé. One hang-glider pilot might be more or less skilled than another and therefore be at more or less risk.

Postcode lottery

Insurance companies try to assess the risk of crimes or accidents more accurately than by just taking average numbers for the entire population. They use complex calculations to work out who is at more or less risk than someone else. This is why your postcode or zip code affects how much you have to pay for house insurance, car insurance, and so on. If there are lots of break-ins in your area, they will assess your house to be at high risk of a break-in and will charge you more.

Increased and decreased risk

A common way of showing risks is by comparing them in terms of factors or percentages. This can be very persuasive, but if we can't see any absolute numbers it is easy to be misled. A statement such as: 'Taking a health pill halves your risk of getting cancer of the toenail,' makes the supplement sound like a good buy. But if the chances of getting cancer of the toenail are only 1 in 20 million, then halving the risk to 1 in 40 million is not really worth the cost of the supplement. You are more likely to have an accident going to buy the health supplement than you are to get toenail cancer.

Some risks can't be measured with the degree of accuracy we might like. If we tried to predict your personal risk of being killed in a traffic accident based on your previous experience, the chance would be zero as you have never been killed in a road accident despite many years of being on the roads.

There are two common ways of misunderstanding risk, that can be summed up in statements like these:

'I've been doing this for years and it's never gone wrong, so I'm sure it will be fine.'

'You've been lucky so far – your luck's bound to run out.'

In a sense, the first one is a sort of fuzzy Bayesian assessment. If we don't know a statistical risk, we make an assessment based on a prior sample. It's not a good idea, though, especially when dealing with risk of death. Of course you have been okay on previous occasions, because you are not dead. Using the previous occasions on which you did not die, you could justify absolutely any reckless behaviour because you have never died on any previous occasion when you did something risky. You won't die this time, because you didn't die last time. But you can die this time, *only because* you didn't die last time.

The second one is also wrong in many instances. This is the inverse of the gambler who keeps betting on the same number because it is bound to come up sooner or later. It's not. Every time, the chance of that number coming up is the same, irrespective of whether it has ever come up before. If you roll a die, there is a 1 in 6 chance of getting a six. If you roll the die and get a six, there is still a 1 in 6 chance of rolling a six next time. So in the case of independent risks, the fact that someone has 'got away with it' for years does not mean they will (or won't) carry on getting away with it.

CHAPTER 23

How much mathematics does nature know?

Can the natural world count?

The medieval mathematician Fibonacci discovered that there is a sequence of numbers which lies behind a lot of phenomena in nature, including rabbit procreation.

Duplicating rabbits

Fibonacci tackled a problem in mathematics that had been known to Indian mathematicians for centuries, but was apparently new to Europe at the time. It goes like this:

If you have two rabbits in a field, how will the population grow, assuming ideal conditions?

The ideal conditions include the following:

- **The first two rabbits are of opposite genders, of breeding age, attracted to each other, healthy and fertile.**
- **Every female rabbit produces a pair of rabbits, one male and one female, every month once she is mature.**
- **It takes a month after conception for rabbits to be born and another month to reach maturity.**
- **None of the rabbits ever dies.**

The last one is taking the ideal conditions to the extreme, not to mention testing the definition of 'ideal', but never mind. This was all 800 years ago and it's too late to quibble.

So we free the first two rabbits into a field where they breed like, well, rabbits. After one month, there is still only the first breeding pair, but they have just had their first babies so it will all kick off soon.

At the end of the next month, there are two pairs: the first one and their now-adult babies. The first pair have another couple of babies, and the second pair get started on their breeding career.

The next month there are three pairs: the originals, the first set of babies and the second set of babies.

Next month, the original pair and their first set of babies both have a pair of babies (still immature) and the second set are ready to start breeding. The set of rabbits grows like this:

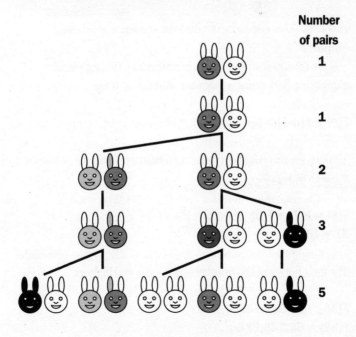

Number of pairs

1

1

2

3

5

And so on. The number of pairs each month follows this pattern:

1, 1, 2, 3, 5, 8, 13, 21, 34 . . .

These numbers don't at first look very interesting, but they crop up again and again. It might not be immediately obvious that there is a pattern to them, but there is. Add together the last two numbers in the sequence to get the next one:

1 + 1 = 2
1 + 2 = 3
2 + 3 = 5
3 + 5 = 8
5 + 8 = 13
8 + 13 = 21

and so on. This sequence is called the sequence of Fibonacci numbers.

If we refer to the nth Fibonacci number as $F(n)$, a general expression for finding a Fibonacci number is then:

$F(n) = F(n\text{-}1) + F(n\text{-}2)$

You can see how this works with an example from the sequence, the eighth number:

$F(8) = F(7) + F(6)$
21 = 13 + 8

The gaps between the numbers get bigger and bigger:

$F(38) = 39,088,169$
$F(39) = 63,245,986$

So

F(40) = 39,088,169 + 63,245,986 = 102,334,155

The numbers escalate rapidly; F(20,000,000) has more than 4 million digits.

If we assumed that Fibonacci put his first two rabbits in the field 800 years ago, and excuse the fact that some rabbits are now 800 years old, there would have been 800 x 12 = 9,600 months for rabbits to grow. F(9600) is more than 2,000 digits long, so greater than $10^{2,000}$. That means there would be more than 10^{20} googol pairs of rabbits by now, or way more rabbits than there are atoms in the universe.

Two bee or not two bee?

The rabbits were all a bit hypothetical, but some other species demonstrate a more accurate representation of the Fibonacci series. If we look at honeybee genetics, the Fibonacci sequence shows the number of ancestors belonging to each one. A male honeybee has only one parent, a female, as it hatches from an unfertilized egg. Each female has two parents, a male and a female. So if you start with a male and draw a family tree, it looks like the one below.

Adding up the ancestors, we get:

	Parents	Grand-parents	Great-grand-parents	Great-great grand-parents	Great-great-great grand-parents
Male bee	1	2	3	5	8
Female bee	2	3	5	8	13

Though the female has a head start, she is only a bit further along the Fibonacci sequence – the numbers are ultimately the same.

Branching out

Many plants grow leaves or branches in a pattern which follows the Fibonacci sequence. It's easy to see why branches fall into this pattern, as each shoot grows a sideshoot and then after a certain point the sideshoot grows its own sideshoots, and so on:

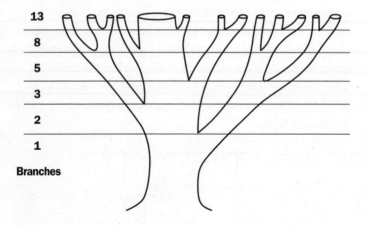

Flowers have Fibonacci numbers of petals, too, and the interiors of most fruits are divided into a Fibonacci number of sections (such as three in a banana and five in an apple). The sequence even appears in our own bodies, as the ratio between the lengths of the bones in the fingers, for example.

CHAPTER 24

Is there a perfect shape?

A quick look around the natural world shows a lot of odd shapes – and some rather elegant ones.

Both the Fibonacci sequence and fractals produce patterns that look less organized than they actually are. Hidden mathematical patterns turn up in other forms, too.

Rectangles and spirals

Try this exercise to see some numbers falling into a pattern. Begin with a square of one unit (let's call it cm, but it could be anything). Draw an identical square alongside it. Now use the two adjacent sides to form the side of a new square (a square with sides of 2cm). You now have adjacent sides adding up to 3cm; draw another square. Keep going until you run out of paper or enthusiasm.

What do you notice about the lengths of the sides of the squares?

1, 1, 2, 3, 5, 8, 13 . . .

It is the Fibonacci sequence again.

Now create a spiral by drawing a curved line that goes diagonally through the squares in sequence.

This is called the Golden Spiral (see top of facing page). Many plants grow their leaves in a golden spiral, coming out of the stem at different angles. The arrangement of leaves on a plant is called *phyllotaxy* and is of great interest to botanists.

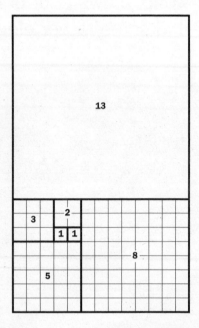

Counting the number of times you circle the stem before finding a leaf vertically above the one you started with gives both a Fibonacci number of revolutions around the stem and a Fibonacci number of intervening leaves (see illustration below). This pattern maximizes the sunlight that falls on any particular leaf, which is why it is so prevalent. The angle between leaves is usually close to 137.5°.degrees.

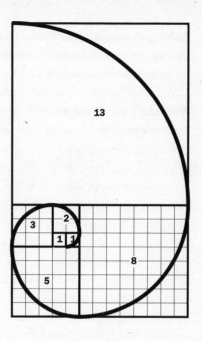

Spiralling spirals

Quite often, several golden spirals are woven together. The seeds in many flowerheads are arranged in patterns of interleaved golden spirals, and the nobbles on a pine cone are arranged in two interlocking golden spirals.

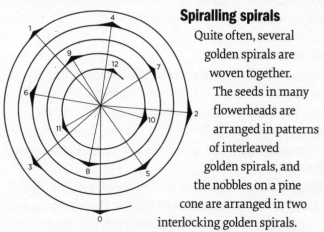

A sunflower has Fibonacci numbers of spirals going in each direction (clockwise and anticlockwise), and the total number of seeds is also a Fibonacci number. This is the best space-packing arrangement, so that way the sunflower can optimize the number of seeds it can fit on a round seed head.

Perhaps the cleverest plant of all is the pineapple. The fruit is covered in hexagonal nobbles, each of which is part of three different spirals. There are eight gently sloping rows of nobbles, 13 steeper rows and 21 nearly vertical rows.

The leaves of the pineapple grow in a different Fibonacci sequence, with five spirals of leaves around the stem before vertically aligned leaves occur. There are 13 leaves between each vertically-aligned pair. This means the pineapple has two sets of golden spirals controlled by different hormones, and switches to the right one when it's time to grow a fruit.

Golden rectangles

Rectangles are varied: there are short, fat ones, long, thin ones, and then some really elegant rectangles known as *golden rectangles*. A golden rectangle has sides in the ratio of approximately 1:1.61803. The number 1.61803 . . . is irrational (the decimals keep on going) and it is represented by the Greek letter Phi, Φ.

This isn't just a random irrational number. It was first defined by Euclid around 300BC. Imagine a line that is cut into two parts. One is longer than the other, but it is very precisely 'longer than the other'. The two lines are in a special ratio to each other, called the golden ratio. The line is cut so that the ratio **short bit : long bit** is the same as the ratio **long bit : whole line.**

Put mathematically, imagine the line is cut into parts a and b. The whole length of the line is obviously a + b. For the lines to be in the golden ratio, a:b must be the same as a: a + b

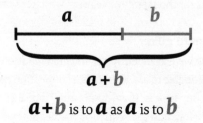

$a + b$ is to a as a is to b

and

$$\frac{a+b}{a} = \frac{a}{b} = \Phi$$

> 'A straight line is said to have been cut in extreme and mean ratio when, as the whole line is to the greater segment, so is the greater to the lesser.'
>
> Euclid, *Elements*

This resolves into a ratio of

$$1 : \frac{1+\sqrt{5}}{2}$$

Chop, but don't change

The golden ratio, and the golden rectangles it defines, turns out to be pretty special. If you take a golden rectangle, like the one on the right, and cut a square from the end (sides a, a), you're left with another golden rectangle (b, a). The sides of the remaining rectangle are also in the ratio 1: Φ. You can keep going, making smaller and smaller golden rectangles.

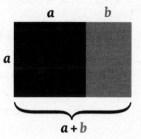

The golden rectangle is generally considered to have the most pleasing proportions. It's found quite a bit in nature, including in our own bodies, and it has been used in artistic and architectural constructions for thousands of years.

Gold, and more gold

Given that we have a golden spiral and golden ratio/rectangle, it would be reasonable to ask if there is a connection between them – and, of course, there is. If we divide any Fibonacci number by the number before it in the sequence, the result tends towards Phi. This is not very obvious at the start:

$$\frac{1}{2} \div \frac{1}{3} = \mathbf{1.5}$$
$$\frac{1}{3} \div \frac{1}{5} = \mathbf{1.667}$$

But using larger and larger Fibonacci numbers, the result becomes closer to Φ:

$$\mathbf{102{,}334{,}155 \div 63{,}245{,}986 = 1.61803}$$

And there's another surprise. If you divide a Fibonacci number by the next in the series, the result tends towards Φ -1:

$$\mathbf{63{,}245{,}986 \div 102{,}334{,}155 = 0.61803}$$

This number – just the decimal part of Φ – is sometimes represented by lower-case phi, φ. From which we conclude that the world has some favourite, lovely shapes.

CALCULATING PHI

Start with $\dfrac{a+b}{a}$

We know that it's the same as a/b and that is the same as Φ.
If a/b = Φ, then clearly b/a = 1/Φ. We can simplify this expression:

$$\frac{a+b}{a} = 1 + \frac{b}{a} = 1 + \frac{1}{\Phi}$$

Therefore,

$$1 + \frac{1}{\Phi} = \Phi$$

Multiplying by Φ gives

$$\Phi + 1 = \Phi^2$$

which can be rearranged to

$$\Phi^2 - \Phi - 1 = 0$$

This is a quadratic equation, so we can use the quadratic formula to solve it for Φ :

$$x = \frac{-b \pm \sqrt{b^2 - 4ac}}{2a}$$

$$\underset{\underset{a}{\uparrow}}{x^2} + \underset{\underset{b}{\uparrow}}{2x} + \underset{\underset{c}{\uparrow}}{1} = 0$$

(a = 1, b = -1, c = -1)

As it's a ratio between positive numbers, we know Φ must be positive, so the solution is:

$$\Phi = \frac{1 + \sqrt{5}}{2} = 1.6180339887\ldots$$

CHAPTER 25

Are the numbers getting out of hand?

Numbers can grow
surprisingly quickly.

There is a legend that the ruler of India was so pleased with the man who had invented the game of chess that he offered him a reward of his own choosing. Even though he could have asked for any amount of riches, the man made what seemed like a very modest request. He asked that the ruler put one grain of rice on the first square of a chessboard, two on the second square, four on the third square, and so on, doubling the quantity as he moved over the board. The ruler granted the request willingly, puzzled that the man should ask for so little. That is, until he tried to present the reward.

The pile of rice grains soon spilled out of the allocated square. It soon overflowed the board, and then the entire palace, and finally all of India. By the time the ruler reached the last square of the board, he needed 2^{63} grains of rice. That's $2 \times 2 \times 2 \times 2 \times$ ... up to 63 times, which is roughly 9,200,000,000,000,000,000 grains of rice. Exactly how much space that would take up depends on the type of rice used. If it was long-grain rice, with grains 7mm (.27in) in length, it would make a line of rice nearly seven light years long. That's most of the way to Alpha Centauri and back, or to and from the Sun 215,000 times.

Exponential growth

Any pattern of growth that depends on increasing by a proportion rather than a fixed amount quickly accelerates. Gabor Zovanyi, a professor in urban planning at Eastern Washington University, claims that if humanity had started with one couple 10,000 years ago, and increased by 1 per cent a year (a bit tricky at the start, but never mind), then by now we would be part of 'a solid ball of flesh many thousand light years in diameter, and expanding with a radial velocity that, neglecting relativity, would be many times faster than the speed of light'. That is not appealing. Fibonacci's sequence of ever-increasing rabbits is another example of exponential population growth, and would get to the solid-ball-of-(furry)-flesh stage a lot faster.

Are we all related?

We can go backwards with population studies, too.

Each person has two parents, four grandparents, eight great-grandparents, and so on, going back through time. Since the number of ancestors increases as powers of 2, it doesn't take very long until you should have more ancestors than there were people on the planet at the time those ancestors would have been living. If we assume 20 years to a generation – which might be a bit short now, but certainly wasn't in the past – you only need to go back to around 1375 to have over 4 billion ancestors. Yet there were only around 380 million people in 1375.

A SELF-FULFILLING PROPHECY

Moore's law, named after the American physicist Gordon Moore, co-founder of Intel, states that the number of transistors on an integrated circuit will double every two years. This is often simplified to say that the processing power of computers will double every two years.

The law, stated in 1965, has been borne out so far, 50 years later. It has become a challenge to the industry, and having a target has helped to see it fulfilled. Moore did not expect it to hold good for more than about ten years.

Somewhere around 1450, there were enough people in the world for each of them to be your ancestor only once – though of course that would not actually have been the case. By 1375, everyone is being more than one of your ancestors at the same time. And they were all being more than one of my ancestors, and your neighbour's ancestors ...

Is that your ancestor on the floor?

As ancestors get reused, there is an increasingly complex network of relationships. This is called 'pedigree collapse' and happens when, for instance, cousins marry, so their offspring have fewer than eight great-grandparents. Pedigree collapse is common in small communities and in elite groups such as royalty.

Yale statistician Joseph Chang has calculated that after a certain point, everyone who lived at that time, and had any descendants, is a common ancestor of all people living in the same community today. That point, for Europe, is around AD600, which means that all non-immigrant Europeans are descended from the Holy Roman emperor Charlemagne (and from lots of other people). The statistical findings have since been confirmed by extensive DNA analysis of Europeans.

Going further back, to 3,400 years ago, everyone who had descendants was a common ancestor of every living person on Earth (in theory). This means that you are related to Queen Nefertiti of Ancient Egypt.

Would you like a loan?

Numbers don't have to double to grow large very quickly.

Proportional increases are familiar to most of us through interest rates. It works in your favour if you are saving, but works against you if you have borrowed money. Banks and financial institutions use a system of compound interest. This means that the interest on a debt or saving is added to the original sum at the end of a period of time (day, month, year) and then the rate of interest is applied to the total.. Suppose you deposited $1,000 at an annual interest rate of 3 per cent. How quickly would it grow?

	Starting balance	Interest	End balance
Year 1	$1,000.00	$30.00	$1,030.00
Year 2	$1,030.00	$30.90	$1,060.90
Year 3	$1,060.90	$31.83	$1,092.73
Year 4	$1,092.73	$32.78	$1,125.51
Year 5	$1,125.51	$33.77	$1,159.27
Year 6	$1,159.27	$34.78	$1,194.05
Year 7	$1,194.05	$35.82	$1,229.87
Year 8	$1,229.87	$36.90	$1,266.77
Year 9	$1,266.77	$38.00	$1,304.77
Year 10	$1,304.77	$39.14	$1,343.92
. . .			
Year 25			$2,093.78

The reason the interest rate matters so much to savers (and borrowers) is that changing it makes a huge difference to these figures:

Capital	Interest rate	10 years	25 years
$1,000	1%	$1,104.62	$1,282.43
$1,000	3%	$1,343.92	$2,093.78
$1,000	5%	$1,628.89	$3,386.35
$1,000	8%	$2,158.92	$6,848.48
$1,000	10%	$2,593.74	$10,834.71

A year at the start is not worth as much as a year at the end. At 10 per cent, the first ten years earn the saver $1,593.74, but the next 15 years don't earn 1.5 times that much – they earn $8,240.97, or around five times as much. This is why politicians and accountants encourage us to start a pension fund early.

Pay for your old age

If you put $1,000 into a pension pot at 20 years old and then retire 45 years later, having paid in no more and gained 3 per cent interest over the period, you would have $3,781.60 when you retired. But if you paid in $1,000 a year for 45 years, still with an interest rate of 3 per cent, you would have $95,501.46 when you retired. If you managed to get 10 per cent, you would have $790,795.32 to retire on, which starts to look respectable – especially for an investment of $45,000.

Day by day

That's fine if you have money to squirrel away – but what if you are at the other end of the economic spectrum? If you have to go to a pay-day lender or loan shark, you can end up paying an astronomical amount in interest because this time the sums work against you.

For example, if you took out a loan for £400 over 30 days at a payday loan rate of 0.78 per cent per day, you would repay £487.36: that's the original £400 plus £87.36 interest. The reason it is so high is that the interest rate – an attractive-looking 0.78 per cent – is charged every day, so every day the capital goes up. The effective interest rate over the whole year is 284 per cent.

What if you just borrowed £50 from a friend for a week and said you'd buy them a coffee? Is that a good deal? It avoids the payday loan. But if a coffee is £2, that's the equivalent of 4 per cent per week – or 208 per cent a year. If you borrowed £50 for a week from the bank at 10 per cent (per year, not per week or per day) it would cost just 10p in interest.

CHAPTER 26

How much have you drunk?

One of the most important tools in mathematics was developed by a German worried about how much he had drunk.

In 1613, the German astronomer and mathematician Johannes Kepler was about to marry his second wife. He ordered a barrel of wine to celebrate. Being canny – and a mathematician – he queried the method the wine merchant used to measure the volume of the wine barrel and so set the price.

Roll out the barrel!

The merchant would put a stick into a hole in the wine barrel, lying on its side, and measure the length of stick that would fit inside. This gave the diameter of the barrel – but at the widest point. The volume calculated from the cross-sectional area of the barrel, multiplied by its height, is an over-estimate of the actual volume, since the barrel is narrower at the ends. Not liking to be cheated either by paying for wine he hadn't had or not having wine he had paid for, Kepler set about finding a better way of measuring the volume of a barrel.

Infinitesimal slices

The method he came up with is called a method of 'infinitesimals'. He imagined the barrel cut into very thin slices, piled on top of one another. Each slice is a cylinder, but with a very tiny height. The cylindrical slices have different cross-sectional areas, with those from the middle of the barrel larger than those at the outside. Of course, each cylinder still has sloping sides, and the round face on one side is very slightly larger than the opposite face. But making the slices very thin, the difference between the two becomes small – infinitesimally small, if they are thin enough – and can be disregarded.

Slippery slopes?

Kepler's method was soon replaced by differential calculus, derived by both Isaac Newton and the German philosopher Gottfried Leibniz in the 17th century.

Newton and Leibniz (separately) were not so much interested in wine as in the slope of a line or curve. They started from infinitesimals: it is clear that the slope of a curve keeps changing, and that you can calculate the slope of any tiny bit of the curve to show a local slope. In the diagram below, making the line **ab** shorter and shorter makes its slope a better and better approximation of the slope of the curve at **a**.

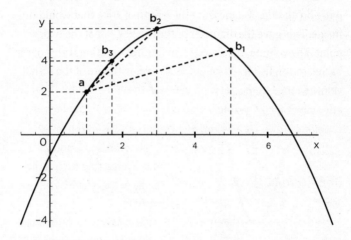

Let's take a simple function, f(2x). The graph of this will be a straight line (see right).

The slope is the same all the way along the line. In fact, the slope is 2 in 1, or 2, because the value on the y-axis (vertical) goes up 2 units for each increase of 1 unit on the x-axis (horizontal) – it is a graph of y = 2x.

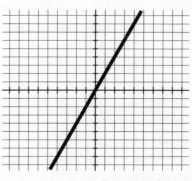

Changing the function by adding a constant doesn't change the slope: the graph for the function f(2x+3) is the same, the line is just at a different point on the axes because now y = 2x + 3 so it's higher up the y-axis (by 3 units) at every point (see below).

Clearly, the constant can be ignored in calculating the slope.

If we now draw a graph of the function f(x²), the graph is a parabola (see below). This has a changing slope. As it happens, at any point on this graph,

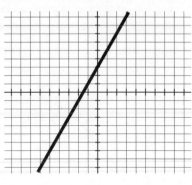

the slope is 2x – as Newton and Leibniz discovered.

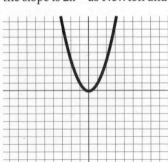

Newton and Leibniz discovered that to find the slope of the graph of f(x) we need to:

(a) multiply each instance of x by its own exponent (power), and

(b) drop the original exponent by 1 in each case.

This is easier to understand with an example. In the function

$$x^3 - x^2 + 4x - 9$$

The exponent of x³ is 3 and the exponent of x² is 2.

x³ becomes 3x² (because we multiply by 3 and drop the exponent by 1, and 3-1 = 2)

x² becomes 2x (because we multiply by 2 and change the exponent to 2-1 = 1)

4x becomes 4 (because we multiply by 1 and change the exponent to 1-1 = 0, so the x only has the value '1' in all cases)

1 disappears: constants (numbers on their own, with no 'x') always disappear as they have no x-exponent.

A general statement of this is xn becomes nx^{n-1}

After all this, $x^3 - x^2 + 4x - 9$ becomes:

$3x^2 - 2x + 4$

This is a powerful result. If we wanted to know the slope at the point where x = 3, we could work it out by substituting 3 for x in the differentiated function:

$f(x^2)$
$f'(x^2)$ is 2x

For x = 3, the slope is 2 x 3 = 6

> ### FUNCTION
>
> A 'function' is any operation that takes input in the form of numbers and produces an output (result). A function is shown as f() with the instructions for the operation in the brackets. So the function $f(x^2)$ means 'take the number x and square it', and the function f(2x) means 'double the number x'.

A single point can't actually have a slope, of course. The slope calculated is that of a tangent drawn to the curve at that point:

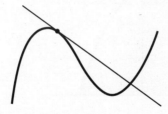

The method is the same even with a complicated function.

$f(x^3 - x^2 + 4x - 9)$
$f'(x^3 - x^2 + 4x - 9)$ is $3x^2 - 2x + 4$

At the point $x = 2$, the slope is $(3 \times 2^2) - (2 \times 2) + 4 = 12$.

Knowing the slope of a graph can give us useful information. On a graph of distance against time for a moving object, for instance, the slope tells us the speed at which the object is travelling. Any function that can be expressed as a ratio or division sum can be related to the slope of a graph. If we plot prices against time, the slope shows the rate of rise or fall in prices (inflation).

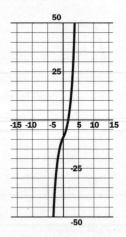

Everything under the curve

While differentiation gives a way of measuring the slope of a curve, *integration* provides a way of calculating the area under a curve. This time, imagine the area beneath the line cut into a myriad tiny columns. By adding together the area of all the rectangles, we can work out the total approximate area.

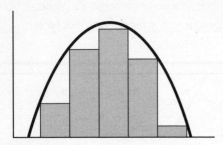

The narrower the rectangles, the better the estimate of the area:

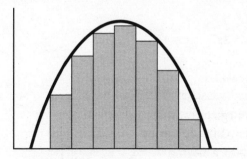

If we could make the slices infinitely thin, we could work out the exact area. This is what integration aims to do.

DIFFERENTIATION

What we call differentiation, Isaac Newton called the 'method of fluxions'. The product of differentiation is called the differentiated function or the derivative. A function of x is written as f(x), and the differentiated function is written f'(x).

Integration is really the opposite of differentiation. If we take the product of differentiation and integrate it, we get the original function (with one small difference).

So differentiating

$$x^3 - x^2 + 4x - 9$$

gives

$$3x^2 - 2x + 4$$

and integrating

$$3x^2 - 2x + 4$$

gives

$$x^3 + x^2 - 4x + c$$

where c is an unknown constant. We can't tell what the constant was in the original function once it's been differentiated.

Integration is just undoing differentiation. You can think of it as anti-differentiation. Differentiating x^n gives nx^{n-1}, so integrating nx^{n-1} gives x^n.

If we want to integrate x^n we reverse the differentiation process: we have to divide by the exponent and raise the exponent by 1:

$1/n \, x^{n+1}$ (because we are undoing nx^{n-1})

This means that the integral of x is $\frac{1}{2}x^2$ and the integral of x^2 is $\frac{1}{3}x^3$. Integration is shown by a long 's', called sigma:

$$\int$$

The statement 'find the integral of $3x^2 - 2x + 4$' is written like this:

$$\int 3x^2 - 2x + 4 \, dx$$

The '*dx*' at the end shows it is the 'x's that are being worked on. If the function used the letter t instead of x, it would end with '*dt*'.

$$\int 3t^2 - 2t + 4 \, dt$$

Integrating

$$\int 3x^2 - 2x + 4 \, dx$$

we get

$$x^3 - x^2 + 4x + c$$
(Don't forget the constant!)

Many graphs can keep going, so they have an infinite area beneath them. We can't calculate the area underneath unless we specify which chunk of graph we're interested in. To do this, we cut it off at two different values of x (or whichever variable we are using).

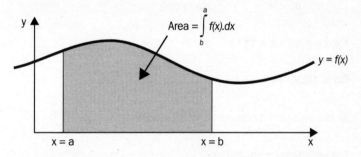

To show which bit we are using, we put the upper and lower limits (that is, the cut-off points) at the top and bottom of the integration symbol:

$$\int_{2}^{5} 2 \, x \, dx$$

This means 'find the area for the curve between x = 2 and x = 5'.

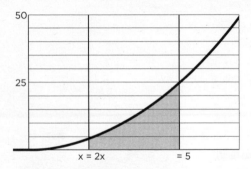

To work it out, we take the result (called the 'integrand'):

$\int 2x\, dx = x^2 + c$

and work it out first for x = 5, then for x = 2, then subtract one from the other (the 'c' will cancel out):

for x = 5, x^2 + c = 25 + c

for x = 2, x^2 + c = 4 + c

So the area under this portion of the graph is

(25+c) − (4+c) = 19

The way out of a hole

Remember the paradox of Achilles and the tortoise? The difficulty arises from dividing time and distance into ever smaller – infinitesimal – portions. Yet this is exactly what differential and integral calculus do. The solution to the problem of this mismatch between the real world, in which areas, lines, volumes and time are continuous, rather than a collection of discrete infinitesimals, came in the 19th century. In 1821, the French mathematician Augustin-Louis Cauchy (facing page) recast the way calculus was presented so that it became just theoretical. Instead of struggling with how to jump the invisible gap between the infinitesimals, he said it wasn't necessary – mathematics is a law unto itself and does not need to mimic or relate to reality.

Perhaps it would be fairer to say that reality does not have to mimic mathematics, since the reality we know is that of continuities and if mathematics fails to model them satisfactorily, that's mathematics' problem, not reality's problem.

> '*[Calculus] is the language God talks.*'
>
> American physicist Richard Feynman (1918–88)

Finally, after 2,300 years, Achilles is allowed to overtake the tortoise.

PICTURE CREDITS